Thought-provoking and practical, grounded in science and real-world experience, Dr. Collins pushes the leadership frontiers. *The Four Stars of Leadership* weaves the insights of Four-Star Generals and Admirals into a powerful leadership system. Whether you're leading a team, an organization, or a nation, this book belongs on your desk.

– Marianne W. Lewis, PhD, Dean and Professor of Management, Carl H. Lindner College of Business, University of Cincinnati

Dr. Tom Collins has pulled together an incredibly comprehensive analysis of the leadership challenges and successes of 51 retired Four-Star Generals and Admirals. These officers managed gigantic organizations with huge budgets. However, the focus is on people. Motivating, training, rewarding, coaching and sometimes inspiring people to do complex jobs in a dangerous and confusing environment. The lessons of *The Four Stars of Leadership* will resonate with others charged with executive responsibility across business, government, academia, and non-profits. Really a fascinating and analytical work.

– General Barry McCaffrey, U.S. Army (Retired), former Commander of U.S. Southern Command

The Four Stars of Leadership is personal, informative, thoughtful, and practical, backed by thousands of years of tested experiences on and off the battlefields of organizations and interpersonal relationships. This is a must-read for anyone who wants to be among the stars when it comes to leadership and each of us making our worlds that much better.

– Barry Posner, coauthor of the bestselling and award-winning book *The Leadership Challenge*

Few books on leadership are as superbly written, well-researched, and thoroughly comprehensive as *The Four Stars of Leadership*. Tom Collins has captured the very essence of leadership in a way that those at every level of the leadership ladder can relate to and learn from. I highly recommend this book for all those beginning their leadership journey, as well as for those well along the way.

– General Anthony C. Zinni, USMC (Retired), former Commander of U.S. Central Command

Not just another "how to" book, in *The Four Stars of Leadership* Tom Collins has crafted a deep dive into what scores of modern, senior military leaders regard as keys to their success. Employing a scientific method, Tom provides a framework of themes and concepts, interwoven with personal context, revealing the traits of wise leadership essential for success in government as well as business. His highly readable text is a valuable addition to library classics such as Emperor Marcus Aurelius' *Meditations* and Governor Zell Miller's *Corps Values*.

– General Robert Magnus, USMC (Retired), 30th Assistant Commandant of the Marine Corps

Powerful – where else can you glean insights, in one place, of over 1,900 years of collective leadership experience from some of the finest leaders our Military has ever produced? No matter what your chosen field, one day you will be called upon to step into a leadership role. Whether you are new to the study of leadership or a long-time practitioner, *The Four Stars of Leadership* has valuable and practical insights to offer.

– General John (Mike) Murray, U.S. Army (Retired), former Commanding General of U.S. Army Futures Command

The Four Stars of Leadership is an exceptional resource for any leader's library crafted from the unique perspectives of 51 retired Four-Star Generals and Admirals. Dr. Collins masterfully curates the insights, experiences, and wisdom into each chapter. Despite leadership's complexity, Tom's logical approach makes it a compelling, engaging read for all levels of leadership experience. Regardless of where you are on your leadership journey you won't want to put it down.

– Admiral Scott Swift, U.S. Navy (Retired), former Commander of the U.S. Pacific Fleet

The Four Stars of Leadership is a masterclass in exemplary leadership and my top candidate for leadership book of the year. Dr. Tom Collins' personal interviews with 51 retired Four-Star Generals and Admirals are unprecedented, revealing, powerful, and illuminating. Nothing like this has been done before, and the real-life stories alone are worth the book's price. They reveal common themes forming the foundation of exemplary leadership applicable to all types of organizations and leaders at all levels. *The Four Stars of Leadership* provides indisputable evidence that the best leadership, inside and outside the military, is not about command-and-control. Instead, it's about character, trust, relationships, moral courage, and selfless commitment to service. What's more, Dr. Collins shares practical lessons on how to put all of this to work in your leadership context. I love this book, and I wholeheartedly recommend you buy it, read it, reread it, and put its lessons into practice.

– Jim Kouzes, coauthor of the bestselling and award-winning book *The Leadership Challenge*

Although senior military officers are expected to be effective leaders, Dr. Tom Collins' insightful and entertainingly instructive work shows both the qualities they share and the personal attributes that make each unique. *The Four Stars of Leadership* is a fun, fascinating read – and a great way to learn.

– General Stan McChrystal, U.S. Army (Retired), former Commander of U.S. and NATO Forces in Afghanistan, and author of the bestselling book *Team of Teams*

Audacious and a unique work of scholarship, *The Four Stars of Leadership* is an essential read for anyone who wants to understand leadership and especially if they want to be a well-regarded, followed leader.

– General Vince Brooks, U.S. Army (Retired), former Commander of U.S. Forces Korea, United Nations Command, and Combined Forces Command

There are hundreds of books about leadership, but if you are to read just one – read *The Four Stars of Leadership*. The author, Tom Collins, has masterfully organized the significant leadership principles and concepts based on his extensive study and interviews of 51 Four Star Generals and Admirals of the United States Military. He weaves their perspectives and experiences into a comprehensive, yet conversational guide on leadership useful to all leaders, young or old, in private or public service.

Leadership is personal. It is a reflection of one's personality, mirroring how we think about life and all the people that impact our lives. In *The Four Stars of Leadership*, Tom Collins has captured how deeply personal great leadership is. Thank you for making it personal...

In *The Four Stars of Leadership*, Tom Collins provides us a look at the most foundational element of the world we live in—leadership. By looking at the experiences of multiple individuals, he gives us a series of insights of value to all.

THE FOUR STARS OF LEADERSHIP

THE
FOUR STARS
OF
LEADERSHIP

Scientifically-Derived from the Experiences of
America's Highest Ranking Leaders.

★ ★ ★ ★

Tom Collins, MD, MS

ISBN:
Hard cover ISBN: 979-8-9926594-0-5
Paperback ISBN: 979-8-9926594-1-2
E-book ISBN: 979-8-9926594-2-9

Cover Design
Julia Kuris

Interior Design and Typesetting
Elisabeth Heissler Design
www. ehgraphicdesign.co.uk

To My Wife

Thank you for being God's hands and feet in my life, for seeing great things in me even when I don't, and for chasing my dreams with me. I am not me without you.

To Anna and Noah

I am a better person because of you. What a wonderful blessing it is to be the dad of two of my heroes! I hope this book will help you to become even greater people and leaders than you already are.

To the Members of the U.S. Military, Past and Present

To those who represent the very best of our nation, who have answered the call to protect and defend the Constitution of the United States against all enemies, foreign and domestic, who have endured the most austere environments and faced the fieriest, hardest crucibles of human experience, who have fought and bled and died or wished they would have, may you dwell in the shelter of the Most High and rest in the shadow of the Almighty.

CONTENTS

FOREWORD

When Dr. Tom Collins reached out to me, requesting an interview to discuss leadership, I was happy to say "yes," as leadership has been my life's work and something I had discussed with others many, many times. Admittedly, I was cautious after reviewing his biographical information. What would a pediatric cardiologist do with the information I would share with him? The answer to my question and perhaps that same question for others interviewed, and you the reader as well, is contained within the following pages.

To be sure, Dr. Collins is audacious. I know of no other work that has interviewed so many life-long leaders. That, achievement, by itself, makes this book a unique work of scholarship. However, the greater achievement is in extracting the essence of leadership, as practiced by leaders who have led repetitively and in increasingly large and complex organizations. Like a master distiller who mixes ingredients and places them over a fire that boils the ingredients, condensing their vapors and cooling them to collect the result into an extraordinary distillate, the author has distilled an extraordinary concoction. The product reflects the lived experiences, from unique individuals each coming from a different place, yet sharing the truly rare, but common in this book, experience of leading through a career in the U.S. military culminating with leadership at the pinnacle of the military leadership enterprise – the Four-Star level.

The output seems so simple in declaring that Four-Star leadership is about Character, Competence, Caring, and Communication. Clear, refined, flavorful. But the simplicity of the distillate belies the complexities of the ingredients.

When Dr. Collins followed up with me after our interview and told me how many Four-Star Generals and Admirals he had successfully interviewed, I was eager to read the work and encouraged him to move to publication, as I have not seen anything like it in my studies of leadership. Moreover, I personally know nearly everyone interviewed, and for the few others, I know well about them.

It is noteworthy that there is so much commonality on the perspectives of leadership. One may reasonably attribute the commonality to "coming up in the same system" and viewing the outcome as the marks of a printing press of sorts, stamping each person as one who made the grade and fit into the mold. This however would be overly simplistic. I would surmise that the "mold" is more the reflection of the collective influence of each of these leaders on the others with whom they served: in senior-subordinate roles, in peer roles, in mentor-protegee roles, and in exemplar-student roles. This is a lattice of people who have shaped and sharpened one another. But it is also reflective of the nature of leading large numbers of human beings, and what it takes to cause them collectively to follow in ways that accomplish seemingly unachievable tasks, under the most arduous conditions with, often, the blood and treasure and reputation of the Nation at stake.

Dr. Collins' recognition of the importance of character is central. These Four-Stars developed character through challenging, difficult situations. Each of them has had his or her character measured and found abundant, honorable, worthy of emulation – again and again and again. Indeed, there are few, if any other leaders who have faced more scrutiny, evaluation and expectation than the Four-Star leaders of the U.S. military. Too often, they are inaccurately associated with power and authority. As Dr. Collins reveals through this work, they are well-acquainted with power, but the use of the power may be surprising for the reader. Instead of the power to rule, which they had, the power is manifested more often among these most-successful leaders as: The power of being the example for others.

The power of competence as the foundation for credibility and trust. The power of including diverse points of view to solve complex problems, adapt and innovate – something familiar to experienced Four-Star leaders. The power of learning and improving as a leader – even when you are hanging on to the top rung. And, the power to develop other leaders to rise to enterprise leadership levels.

The Four-Stars interviewed for this book, and the others who were not interviewed, rise above profit, pettiness or politics, emphasizing honor and character over outcome. The outcomes still happen, and there is among Four-Star leaders an extraordinary list of favorable outcomes, but all would recognize that the outcomes were delivered by the people they led, not by themselves as the leaders.

Enjoy reading this book. Sample the refined distillate of pinnacle leadership that Dr. Tom Collins is offering to you, for truly he has succeeded in examining the hearts of an unprecedented number of leaders, all Four-Star level Generals and Admirals, to distill this essence.

This is an essential read for anyone who wants to understand leadership and especially if they want to be a well-regarded, followed leader.

GENERAL VINCENT K. BROOKS, U.S. Army (Retired)

INTRODUCTION

Anybody can become a good leader. It just takes understanding and persistence. Some people, it might appear, because of their good looks and their way they approach things, were [...] natural born leaders. Well, even those people have to apply themselves, or they're not going to really become consistent, good leaders. Those who aren't born with this special talent can still become equally good leaders

– ADMIRAL JIM HOGG, U.S. NAVY –

Since you're reading an Introduction to a leadership book, I'm going to make two assumptions. First, either you are a leader who wants to be better, or you are someone who wants to be a leader but isn't sure how. Second, you are trying to decide if reading *The Four Stars of Leadership* is going to be worth your time, effort, and money. In response to the first assumption, I say, that's great! We can all be better leaders, and the greatest leaders are always seeking to improve. So, whether you choose to read this leadership book or another one, you are in great company. To the second assumption, I get it. You see the untold numbers of leadership books on the market and wonder something like, "Why is this the leadership book I need to read, rather than some other book?"

That's a fair and important question—one I grappled with throughout the process of conducting interviews for this study. In fact, every one of the 51 retired Four-Star Generals and Admirals who participated in the research for this book contributed insights that shaped its unique perspective. By the time I had spoken to all of them, I realized that their collective wisdom provided the foundation for a unique book on the topic. And so, while I can't directly answer for you the question of whether this book is worth your time, I can help you decide by jumping straight to what differentiates *The Four Stars of Leadership*.While there's a lot that makes this book distinctive, I suspect you have limited time to make your decision. So, to help you do that as quickly as possible, I'm going to do something unconventional—provide bullet points of what you will get out of *The Four Stars of Leadership*. After that, if you're still interested, I introduce more of what the book holds and how it came to be. But, for now, here are the things you can expect from this book.

What Readers Will Get from *The Four Stars of Leadership*

- **A Robust, Yet Simple Leadership Framework**
  I integrate scientific rigor, diverse perspectives, and first-hand practical insights from some of the highest-ranking and most experienced leaders in U.S. history

- **Powerful Stories**
  I provide illustrative examples, many of which will leave you speechless, from leaders who played roles in globally significant events, adding depth to the lessons conveyed

- **Universal Applicability**
  I lay out leadership principles that are relevant across various fields beyond the military, including business, healthcare, academia, and community leadership—nearly all the interviewees have excelled in high-level roles outside the military, such as CEOs, board members, university presidents, and government officials

- **A Comprehensive Leadership Guide**
  I address multiple facets rather than focusing on one area or approach, as most other leadership books do

- **Science-Based Insights**
  I employ rigorous, scientific methods, including conducting interviews with 51 retired Four-Star Generals and Admirals, offering empirically grounded leadership insights

- **An Unrivaled Collection of Expertise**
  I harness the collective experience of nearly 2,000 years of military leadership, providing an unparalleled leadership resource

- **Expansive Perspective**
  I detail a wide range of real-world experiences and leadership challenges faced by a diverse group of dozens of highly accomplished leaders, as opposed to the isolated anecdotes and singular perspective of one author or interviewee

- **A Focus on High-Impact Leadership Roles**
  I reveal practical lessons from leaders who led immense commands, managed incredible budgets, and held influential positions on the world stage

- **A Candid Depiction of Military Leadership**
  I clarify the misconception around military leadership as purely "command-and-control," revealing the trust, relationships, and complex leadership skills required

Now that you know what you'll get from *The Four Stars of Leadership*, I want to fill that out a bit more fully. To do that, let's start with why we need another leadership book. From there, I'll cover why we need *this particular* leadership book. I'll go on to describe how *The Four Stars of Leadership* came to be and share what qualifies me to write such a book. Finally, I'll close out this Introduction with some practical points about the way some of the information is presented in the book.

WHY DO WE NEED ANOTHER BOOK ON LEADERSHIP?

There's no doubt that leadership is an important topic, one with direct impacts on any collaborative human endeavor. It's also clear there are a lot of books addressing various facets of leadership in different domains of activity: leadership in the mountains, leadership at sea, leadership on the football pitch, leadership in the C-suite, leadership to produce profits, leadership for efficiency, leadership to achieve buy-in, leadership to power change, *ad infinitum*. With all those books on leadership, why would we need another one?

Having read hundreds of leadership books and having reviewed innumerable others on the market, it is clear that almost all suffer from one or more weaknesses. First, the large majority concentrate on a single facet of leadership, such as figuring out why your organization exists or instituting change. Others focus on a single objective, while overlooking an array of other things vital to leadership success. I have read untold numbers of books like this, and when I finished, thought, "Wow, that book could've been covered in one sentence." There just aren't a lot of books out there that attempt to cover leadership in a complete way. Second, most are written from a single person's perspective and experience. The authors are often either current or former leaders of companies that figured out how to sell something for a large profit, or consultants who studied those same companies. Perhaps they are stories of how this one guy faced the impending collapse of a given company and was able to turn it all around, and here's how he did it. Since the books are anecdotal, no scientific method or rigor shapes the content, and there is little quantifiable reason to believe the events are reproducible. Additionally, the perspectives on which the books are based almost always flow from one person (i.e. the author) or only a few people, which really limits the veracity and applicability of the resulting ideas or theories. Another shortcoming is most leadership books available are focused on the marketplace as the principal, if not only, domain of leadership, often neglecting leadership in families, communities, sports teams, healthcare, academics, faith organizations, and non-profits, among others. Finally, many of the books talk about management and call it leadership; though both are important, the two are distinctly different. Of those that do focus on leadership, most focus only on a narrow facet of it. They also usually imply, if not declare, that addressing that one particular element will revolutionize a company, family, community, or the world. Yet, our companies, families, communities, and world are suffering for want of leadership. Thus, despite the thousands of leadership-related books on the market, there remains a major need.

HOW DOES THIS BOOK ADDRESS A NEED?

If there is still a need for a book on leadership, how can this book help fill it? I have designed *The Four Stars of Leadership* to address each of the four weaknesses identified above. When I began the project that would culminate in this book, I had an overarching goal: answer the question, "If you were stranded on a desert island with a group of people and could have only one book on leadership, what would that book be?" The book that answers that question has to be comprehensive, but it also has to be manageable—no one will read (or carry around on a desert island) a gigantic, unabridged treatise on leadership. The book needs to be the Swiss Army knife of leadership books. This book attempts to walk that line. It is up to you, the reader, to determine if I have accomplished that goal.

The Four Stars of Leadership is based in science. Specifically, I performed the interviews using rigorous scientific methods. This started with designing the study and obtaining approval from an institutional review board (i.e., Stanford University, protocol #66218). Next, I conducted interviews of 51 retired Four-Star officers using a semi-structured methodology. I will explain why I selected Four-Star officers momentarily, but suffice to say as a group they are among the most experienced and accomplished leaders anywhere. I interviewed *retired* Four-Stars because the Department of Defense requirements make it nearly impossible to interview active Four-Star officers for a project like this, and of course there are also more of them. Each participant answered the same four interview questions.

Interview Questions for the Four-Star Officers

1. What are the three most important leadership lessons you learned across your career? (Follow-up: Can you give an example of how each played out in your career?)
2. What is a leadership lesson you learned the hard way?
3. How do you want to be remembered as a leader?
4. What are five leadership maxims you always try to keep in mind?

Following completion of the interviews, I analyzed the nearly 300,000 words I'd collected from these Generals and Admirals. I did this in a specific way, to be able to identify the major themes they were discussing. There were 115 leadership themes! From those, I identified a subset to be included in *The Four Stars of Leadership*, including those discussed most, those most universal (applicable to the largest number of leaders), and those particularly noteworthy, whether from their novelty or importance.

Unlike most books on leadership that draw from one or a few perspectives, I cover it from the perspective of dozens of extraordinarily accomplished leaders. In fact, no other book has directly or scientifically collected the insights of Four-Star officers like this. These leaders have led across the spectrum, from small groups to millions of personnel, and they bring a host of different leadership perspectives, experiences, successes, and failures from some of the most challenging leadership environments known. That's why I chose them for this work, and it means the leadership principles and lessons in this book are drawn from numerous perspectives, providing a more complete examination of leadership than could be gained from a singular perspective.

While drawn from heralded military leaders, the leadership lessons in *The Four Stars of Leadership* are universally applicable. This isn't a book that provides you with strategies of war. It won't teach you to be a military officer. In fact, while most of the examples are drawn from a military setting, the majority are not from the heart of combat. Many could just as easily have occurred in a corporation, hospital, university, or community. The leadership guidance offered here can help anyone become a better leader, no matter the context.

WHY STUDY GENERALS AND ADMIRALS?

Many readers may be asking how the leadership lessons from the U.S. Military's most accomplished leaders could help them lead better in their homes, communities, and careers. To answer that, I need to address the implicit statement underlying the question and then provide some background on what it means to be a Four-Star General or Admiral.

In my professional domain of academic medicine, I have been told many times, "Military leadership will not work in medicine." When I hear that claim, it tells me a couple of things: the person saying it knows very little about the military and even less about leadership. That may sound harsh. It's meant to be. It's meant to shock people into stopping and, hopefully, questioning their own uninformed conclusions. Those who make such claims may be relying on Hollywood depictions of the military, productions filled with intense moments where immediate action is critical for survival, and there is no room for error and no time for dialogue or rebuttal. As a result, they have extrapolated those "command-and-control" moments to all military activities, never recognizing all the preceding leadership of an entirely different type that it took to establish trust and relationships—what the military terms "cohesion." They don't realize it is someone's prior experience with their leader that forges the relationship to allow them to entrust their survival to their leader in those life-and-death moments. Contrary to the belief of many people, including my colleagues in academia, no soldier risks their life to storm a bunker because they are afraid of punishment from their superior for disobeying an order. They do it because of the relationship. I use a thought experiment to try to help my colleagues recognize this.

Thought Experiment About Military Leadership

Imagine you're a soldier and your commanding officer instructs you to deliver a package to a unit in a nearby building. Suppose there is a street between your building and your destination. Now, imagine there are dozens of trained snipers with their weapons fixed on that street with the order to kill anyone who attempts to cross it. What would make you pick up the package and attempt to cross the street? Would fear of your boss do it? What about fear of losing your job? Would fear of going to prison make you attempt to cross the street?

It's hard to imagine anyone would be so afraid of the consequences of disobeying the order in the thought experiment that they would choose a nearly certain death at the hands of one or more enemy snipers. If you have no military experience, perhaps following that thought experiment, you may be willing to consider that command-and-control is not the form of leadership that drives the U.S. Armed Forces. However, you may have no context for knowing what it means to be a Four-Star General and Admiral and may be wondering what sort of leadership expertise a Four-Star might have. Effectively, you may be asking yourself, "Why study what Four-Star Generals and Admirals have to say about leadership?"

The Continental Army, established in June 1775, was the genesis of the United States Military. That means, as of the time of this writing, the U.S. Military has been in existence for 249 years—nearly a quarter of a millennium. Across that time, the military has preserved the Constitution, fought and defeated tyrants, provided rescue and alleviated pain and suffering during natural disasters, advanced countless now-commonplace technologies, cleaned up environmental catastrophes caused by corporate entities' negligence, stood in the breach to ensure all citizens' rights were upheld, and trained the first humans to land on the Moon. Throughout that exceptionally accomplished history, the military has been devoted to developing and training leaders, and those leaders have been instrumental in the success of the United States of America. Excluding the special circumstances of World War II, the highest rank in the U.S. Military is Four-Star General (in the Air Force, Army, Marine Corps, and Space Force) or Admiral (the equivalent rank in the Navy and Coast Guard). There have been fewer than 900 Four-Star officers in U.S. history, and there are currently approximately 250 living, retired Four-Star officers. Thus, the 51 Four-Stars who participated in this project represent a significant collection of the deep leadership knowledge and training of one of history's most successful organizations.

No Four-Star General or Admiral starts out at the top. They all begin at the bottom of the hierarchy, typically as a second lieutenant or ensign, though some are enlisted personnel and later are commissioned as officers at those ranks. Among those newly minted officers, less than 0.03% will reach the rank of Four-Star. For those who do, it takes decades-long careers of accomplishment in which they are trained and serve as leaders. If they perform well, they are promoted to a higher rank with greater responsibility for increasingly larger groups. Only after over 30 years of development and service, if they have been exemplary along the way, have received multiple congressional confirmations, and have been selected by a board of Four-Star officers, will they reach the rank of Four-Star. This is to say that Four-Star officers are extraordinarily bright, trained, and accomplished leaders. They are the standard-bearers of the U.S. Armed Forces.

The group of Four-Stars I was honored to interview for *The Four Stars of Leadership* are as extraordinary as their peers. Among them are 48 men and three women, including the first female U.S. Four-Star officer, the first female Air Force Four-Star General, and the first female Four-Star Combatant Commander. Their leadership experience is enormous, with an average length of military service of 38 years, combining to a total of 1,941 years of military leadership experience (and not including all of their leadership experience after retiring from the military). These Four-Stars have led groups of size and financial scale that are inconceivable for many leaders. At the pinnacle of their careers, the organizations they led had a median size of 140,000 personnel, with the largest comprised of over 3 million! They also had median annual operating budgets of $12 billion, with the largest totaling over $770 billion! This does not include the value of assets like bases, aircraft, ships, etc., which in the case of the Commander of the U.S. Navy's Pacific Fleet is worth half-a-trillion dollars. Many among my interviewees have further distinguished themselves, as 29 served on the Joint Chiefs of Staff (with three also serving as Chairman, the highest-ranking member of the U.S. Military). Nearly all hold graduate degrees, with about half having multiple degrees, and three hold doctoral degrees. Further, following their military service, almost all have served on the Board of Directors of corporations: 14 have served as Chairman of the Board, and 15 have served as CEOs. The majority serve as consultants to Fortune 500 companies. Four have served as

University Presidents, and others have held distinguished leadership positions such as Secretary of Defense, Director of the CIA, Secretary of Homeland Security, National Security Advisor, Ambassador to China, and Special Envoy for Middle East Peace. It would be tremendously difficult to find a group with greater breadth and depth of leadership knowledge and experience! All of that has been synthesized into the book you are holding, along with first-hand stories to illustrate key lessons. Some of those stories are a part of events that changed world history, and you have the opportunity to learn from them.

WHAT GIVES ME THE AUTHORITY TO WRITE THIS BOOK?

In March of 2018, I was in the throes of building a highly successful clinical program at Stanford University. I'm a pediatric cardiologist with special expertise in caring for children with genetic conditions that cause abnormalities of their heart and arteries. The program I was building brought together a large group of specialists to care for those kids because their conditions often affect many parts of their bodies. In the midst of the process, I realized I needed to commit myself to becoming a better leader in order for the program to reach its full potential, as well as for me to reach my own.

The program at Stanford wasn't my first foray in leading a group. I had built and led multiple other successful clinical programs. Additionally, I had received leadership training and experience through over a decade of service in the Air National Guard, where I achieved the rank of lieutenant colonel and served as the Chief of Clinical Services for our medical group. My prior training and experiences gave me enough capability to do a good job in building the program at Stanford. More importantly, they equipped me with the ability to recognize that I needed to continue to grow and develop as a leader if the program and I were to achieve what was possible. Consequently, I began dedicating my personal reading time to the study of leadership.

After 18 months of dedicated study, I had read about 140 of the most popular leadership books and felt like I had a pretty good grasp on what was out there. However, I also suspected there were probably some unknown unknowns—things I didn't know I didn't know. So, I enrolled in a master's program at The Citadel to study leadership, graduating in 2022. That program provided a formal framework for me to think about leadership practically, conceptually, and scientifically. It inspired me to conceive and conduct studies on leadership, some of which I have published, and others I'm working on publishing. Moreso, when combined with the now hundreds of books I've read on leadership, my master's program gave me the idea for the leadership book I really wanted to read.

As I thought about the leadership book that I wanted to read, several features jumped to mind. It would be a book that explored the multifaceted nature of leadership yet provided a simple framework to think about and develop excellent leadership in any domain. It would be built on the perspectives of numerous exceptional leaders, specifically, Four-Star Generals and Admirals from the United States Military—people who, as a former military member, I looked up to as exceptional leaders. Those leaders would bring to bear their vast experience in the most trying of leadership circumstances. Those experiences, both good and bad, would translate into real world examples of powerful leadership lessons. Further, the book would distill numerous leadership concepts into concise, portable principles, many of them presented in figures that are easy to understand and remember.

I knew the book I wanted to read, but amongst the thousands of leadership books available, it didn't seem to exist. I realized that since I knew the literature, had the writing ability, and most of all had the idea, the responsibility fell on me to write it. Then reality set in—I didn't know any Four-Star Generals or Admirals. How does someone go about contacting a bunch of Four-Star officers? Afterall, these are the highest ranking, most accomplished officers in the military; it's not like their information is readily available, and why would they take the time to talk to me, even if I could reach them? I had a major logistical hurdle to overcome if this book would ever come to fruition.

HOW DID I GET CONNECTED WITH ALL THESE FOUR-STARS?

Because of their high visibility and backgrounds, many Four-Star Generals and Admirals make it a point to be difficult to find and contact.

After hitting multiple dead ends, I decided that my book probably wasn't going to happen. Then, in the Spring of 2022, after having completed a scientific study that distinguished between the concepts of leadership and management, I had the opportunity to meet with one of the study participants, Lieutenant General Joe DiSalvo. During our conversation, I mentioned to him my idea for studying the leadership wisdom of Four-Star officers. Though he didn't have any contacts with Four-Stars, which says something about the rarefied air of the rank, he gave me some ideas for how I might go about contacting some, such as contacting the Pentagon, using LinkedIn, and through the Hoover Institute at Stanford where I was at the time. Thinking the odds were low, I nevertheless decided to try his ideas because I didn't have anything to lose.

To my amazement, I was quickly able to get in contact with multiple Four-Star Generals. Thereafter, I started the interviews. At the end of each, I would ask them if they could connect me with five other Four-Stars. Many were gracious enough to assist me in that regard, which allowed me to develop a "cascade." In less than nine months, I went from having no idea how to contact one Four-Star to having interviewed 52 (one was off-the-record and isn't included in the book). I still marvel that this all occurred.

How Is This Book Organized?

As someone who has taught various subjects, from chemistry to medicine, leadership to career development, I have made every attempt to organize the material in *The Four Stars of Leadership* so that readers can learn, remember, and apply it easily; that is, so that they can become better leaders through improvement in numerous areas. This goal informed how I conducted the study, analyzed the data, assigned terms to themes, and organized the book.

During the process of analyzing the transcripts, I identified four major themes that are based on the perspectives of the Four-Stars, irreducible with regard to their role in leadership: Character, Competence, Caring, and Communication. These four themes came out in the interviews of all of the Four-Stars, and they are ordered based on the frequency with which they were discussed and the importance attributed to them by the interviewees. These themes became the "four Cs of leadership," which then became the "four stars of leadership." Consequently, the book is organized into four distinct sections that address each of these four leadership topics.

The format for each section is similar. The first chapter is an introduction to the principal theme of the section (i.e., Character, Competence, Caring, or Communication). Following that, there are chapters covering specific leadership concepts that fall under the section theme. Each chapter opens with a personal story from one of the Four-Stars that ties directly to the main topic of the chapter. A discussion follows of that leadership topic, synthesized from direct quotes from the interviewees, leadership literature, and my experience, training, research, and writings in leadership.

SOME COMMENTS ABOUT HOW THE BOOK IS WRITTEN

Recognizing that much of my prior writing and publications has been academic, with *The Four Stars of Leadership* I wanted to produce an approachable and useful guidebook, not an academic product. Consequently, I have tried to write it in more of a conversational voice. To help with this, I have employed personal pronouns and have made the leap to assume you recognize you are a leader. I have also chosen to provide fewer citations than I typically do. There are myriad academic papers and books related to the topics of discussion contained within this book. Instead of inundating you, I have attempted to note only those works that are either particularly salient or provide scientific support for concepts that some readers might question and have included them in an Appendix instead of in-text citations.

The figures within the book are all original and are meant to convey the principal concept being discussed. Where there are graphs, these are not intended to convey scientifically acquired data, but rather are conceptualizations of the relationships between particular concepts derived from my synthesis of all the data I collected, the literature, and my expertise on leadership. For those who would seek to refute my conceptualizations, I'm happy to have a conversation. For those who would like to investigate their accuracy, let's collaborate!

With regard to quotations, there are a couple of things worth noting. Each Four-Star—with the exception of Admiral Leighton Smith, who passed away before the project was complete—reviewed all sections of the manuscript in which they are quoted or mentioned and have approved my usage (Admiral Smith's family members reviewed his posthumously). As much as possible, I want the Four-Stars to speak for themselves and have provided as much direct, unadulterated, and lengthy quotes as reasonable. To accomplish this, I have employed a couple of tools: the ellipsis and brackets. I have primarily used the ellipsis to show that some amount of material has been removed from the quotation. This is often when the speaker has taken a sidebar, reiterated words, or has had false starts to a sentence or phrase. The goal with removing them is to provide concision without losing any of the substance of the speaker's quote. Occasionally, I have used the ellipsis at the end of a quotation to convey a speaker's pause due to the emotional weight of a recollection or their sincerity regarding the topic. When you see brackets, one of a few things is happening. The speaker may have used a form of a word that might've been appropriate when a sidebar was present and was rendered unfitting when an ellipsis was employed, or the original word would be ambiguous to the reader due to a lack of preceding contextual information. For instance, a speaker might've previously said, "Communication is crucial" and gone on to discuss it in more detail, returning to say, "It is the most important thing in leadership." For the sake of clarity and streamlining the idea, this might be edited to be "[Communication] is the most important thing in leadership." Second, the speaker may have spoken about something or someone and used a truncated term or title that may be unclear for some readers. For instance, a speaker may have said, "Rumsfeld called me and told me to come to his office." For clarity, this would be edited to "[Secretary of Defense Donald] Rumsfeld called me and told me to come to his office." Finally, military members have a reputation for using profanity. Because profanity can be distracting for some readers, in many cases I have elected to either use brackets and replace the word with a reasonable synonym or have employed an ellipsis. In the situations where I have retained the profanity, it has been because it demonstrates the speaker's emphasis or passion about the subject and wasn't so severe as to derail the subject.

There is some terminology that is worth clarifying, specifically around the use of two words: four and stars. When you see "the Four-Stars," this capitalized and hyphenated term is in reference to the military officers who participated in the project. When unhyphenated, as in "the Four Stars of leadership," this is in reference to Character, Competence, Caring, and Communication. "Four-Star leadership" is the pinnacle of leadership, what we should be striving to reach. It is the type of leadership that is comprised of excellence in the "Four Stars of leadership" and is that level of leadership demonstrated by most of "the Four-Stars" who participated in the project.

As the quote from Admiral Jim Hogg states at the beginning of this chapter, anyone can become a good leader if they devote themselves to it. With a commitment to self-improvement, the right resources, and the opportunity to enact their learned leadership skills, everyone can become a better leader. *The Four Stars of Leadership* offers a simple framework for conceptualizing, learning about, and developing leadership expertise. In this book, you have a rich cache of leadership stories, lessons, and wisdom from some of the most accomplished leaders in history. If you want to be a Four-Star leader of the highest caliber, read on.

CHARACTER

THE FIRST STAR OF LEADERSHIP

CHAPTER 1

CHARACTER:
THE FOUNDATION OF LEADERSHIP

Leadership is a potent combination of strategy and character.
But if you must be without one, be without the strategy.

– GENERAL H. NORMAN SCHWARZKOPF JR., U.S. ARMY –

When General Charles Krulak was Commandant of the U.S. Marine Corps—the highest-ranking Marine—he and his wife, with the help of others, enjoyed baking cookies to take to the Marines who were on-duty on Christmas day. The cookie baking was a huge operation; they baked hundreds and hundreds of dozens of cookies. They put the cookies on paper plates, and General Krulak would place a Marine Commandant's card on each. They would then load the cookies into a van and take them to every duty post in the Washington, DC and Quantico, VA area. General Krulak would make his way to every duty station in each barracks, on each floor, at each fire-watch on each floor, until he had visited every on-duty Marine, personally handed them a plate of cookies, and thanked them for their service.

On Christmas day in 1998, General Krulak went about delivering the cookies to the on-duty Marines. Having finished visiting all the duty posts in the Washington, DC area, General Krulak made his way

★

to Quantico, where he visited the barracks at the Officer Candidate School and the Basic School. He then visited all remaining barracks before finally stopping at the headquarters of the Marine Corps Combat Development Command.

When General Krulak walked into the duty hut at the headquarters, the staff sergeant on duty snapped to attention. Krulak greeted the Marine and said, "Here you go. I've got some cookies for you." After exchanging gratitude to each other, General Krulak said to the young staff sergeant, "And I have some for the officer of the day. Is he here?"

The Marine replied, "No, sorry. He's out walking the post, seeing what's going on."

Krulak said, "Oh, that's good. That's what he should be doing. Who is it?"

Unhesitating, the staff sergeant said, "Brigadier General Mattis."

Krulak, thinking he must've misunderstood the answer said, "What? Who is it?"

The staff sergeant again replied, "Brigadier General Mattis."

General Krulak now saw that the young Marine must've misunderstood his question and said, "No, I don't mean who's on duty and is at home and on call. I mean, who's here working with you?"

Resolutely, the staff sergeant again replied, "Sir, it's General Mattis."

General Krulak, becoming irritated, looked the sergeant in the eye and pointedly said, "Okay, let me cut to the chase here. See that cot over there? Who's going to sleep in that cot tonight?"

Containing his own frustration, the staff sergeant returned General Krulak's gaze and repeated, "Sir, it's Brigadier General Mattis."

Amazed at the situation, Krulak was attempting to figure out how to resolve the miscommunication when he heard a sound—"clink, clink, clink, clink." He recognized it as the sound of a sword striking someone's leg as they walked, and around the corner walked Brigadier General Jim Mattis in his dress blues with his sword at his side.

He was, in fact, the duty officer.

Looking at Mattis with surprise, General Krulak said, "Jim, with all due respect, what the hell are you doing standing your duty? It's supposed to be a lieutenant or a captain, and you're here."

★ ☆ ☆ ☆ ☆

Mattis matter-of-factly replied, "Well, I'm a bachelor."

"I know you're a bachelor," replied Krulak. "What's that got to do with it?"

"Well, I looked at the duty roster," explained Mattis, "and I saw that it was a captain who had two kids. And I said, 'Hey, why should I have Christmas Eve and Christmas Day off while this captain who's got a family isn't going to be there with them?'"

As General Krulak relayed the story, nearly 25 years later, it was clear he remained amazed and inspired by Mattis' character and action, reiterating: "...he didn't call anybody. He didn't tell anybody. I [was] the Commandant. I would've never known. I'm sure that [Commanding General] McKissock would've never known if I hadn't bumped into him. That's the kind of guy Jim Mattis is." For General Krulak, what he saw in Jim Mattis that day demonstrated what he sees as the foundational component of leadership. Krulak explained:

> There is only one "maxim" that stands above all others [...] one trait that is foundational if one is to be an effective leader in the military, industry, or personal life. It cannot really be taught in the typical sense of the word [...] it must be seen, recognized, understood, appreciated, and mirrored. That foundational trait is that the individual be a man or woman of character. [...] Give me a person of character, and I will show you a leader.

There's no doubt General Krulak views character as the fundamental attribute to effective leadership. He isn't the only one. In fact, all 51 Four-Stars either explicitly stated the essential role character plays in leadership, or they discussed the vital contributions of numerous components of character to leadership. General Bob Kehler, former Commander of United States Strategic Command, echoed Krulak's emphasis on character, saying it "is the most important attribute for a leader." General Kehler put character into perspective by explaining it establishes "the credibility of a leader who may have to order people to do something very dangerous or [with a] very high likelihood that might get them killed. We don't give that responsibility to people who lie, cheat, steal, or tolerate those who do, nor should we." But if character is so fundamental to successful leadership, and it is composed of multiple elements, this raises a crucial question.

★

WHAT IS CHARACTER?

Most people recognize the depth of character of someone who, as a One-Star General, would voluntarily take a captain's duty on Christmas day. It's one of those things that is fairly easy to recognize when we see it—or don't see it—but it can sometimes be more difficult to define. That doesn't help you if you are hoping to become a more effective leader by improving your character. So, what is meant by character; how can it be defined?

A person's character is composed of a combination of virtues and the degree to which those are manifest—acted upon or brought to life—in someone. It is more than an ethereal cloud of goodness or morals that surrounds someone; it is a state of active expression. The degree to which someone actively expresses character ranges across a spectrum from poor or questionable to exemplary or unassailable. Typically, when someone talks about a person of character, they are expressing an implicit, positive value judgment; they mean a person of high moral character. So, the question of character is really more what constitutes high moral character; what are the components that form it?

A composite of virtues

Many of the Four-Stars spoke directly about the importance of character as they talked about leadership, and all emphasized the imperative of various personal virtues that compose it. Often character is something we think of as a single attribute, like the kindness shown in the thoughtful act of General Mattis above. However, in reality, character is composed of multiple virtues. For instance, Admiral James Loy, who served as the 21st Commandant of the U.S. Coast Guard, as well as Secretary of the Department of Homeland Defense said, "honesty and integrity, courage, respect, commitment, trust, [and] ethics" are "foundational values of one's character." This dual nature of character demonstrates its complexity: it is a unified and indivisible expression of one's virtue, while at the same time being a composite of the virtues one possesses.

★☆☆☆

The composite virtues of character can be segregated based on the elements of one's character. As shown in Figure 1.1, those virtues separate into two types: A) Those principally related to oneself and B) Those principally related to others. For instance, integrity is alignment—complete harmony—among what you believe, what you say, and what you do. It is self-related and expressible apart from another person. Conversely, there are those virtues that are related to other people, such as honesty or compassion.

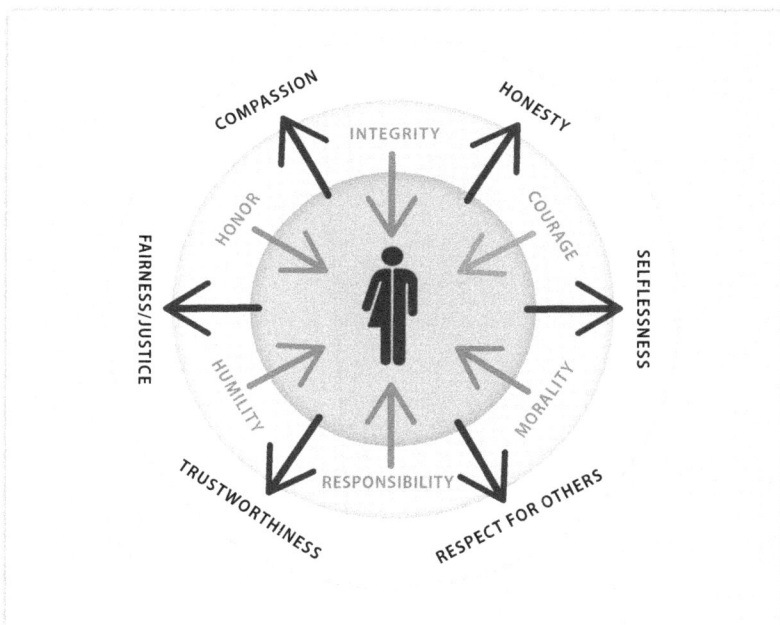

Figure 1.1. The Most Commonly Cited Virtues Comprising Character. The Four-Stars discussed multiple virtues that constitute a person's character. These virtues can be segregated into those that are personal, inward-directed (gray arrows) and those that are social, outward-directed (black arrows).

Thinking of virtues in a dichotomous manner—inward-focused versus outward-focused—may be new to many readers. I suspect that for most people, the tendency has been to lump the different virtues into the one big amalgam we call "character." But, separating them into the two groupings can help us begin to understand character in a more nuanced way. It gives us a framework to be able to identify specific aspects of our character that need improvement. For some of us, it may be that we have issues with our personal, inward-directed character virtues. Perhaps we are fair and honest, but we lack humility or personal responsibility. Conversely, someone else may have issues with those social, outward-directed virtues. An example here might be someone who is marked by courage and responsibility but lacks compassion and respect for others.

While it is valuable to recognize the dichotomy and independence of the virtues, they all remain related and interconnected. For instance, if we consider honesty, which we often think of with regard to being honest with others, there is a personal side to honesty, namely "being honest with yourself." That is more accurately an issue of integrity, but it also demonstrates the interdependence of virtues. Another example would be the relationship between courage and selflessness. The ultimate selflessness is to put oneself in harm's way, risking life and limb, and courage underlies the ability to do so. Similarly, most all virtues, both self-related and others-related, are tightly connected. This leads to yet another question: Is character an "all-or-nothing" proposition; do you either have all the virtues or none?

HOW DOES CHARACTER MANIFEST?

Generally, in society, there seems to be an expectation or demand for 100% manifestation or mastery, all of the time, of all the virtues that compose character, especially for those in leadership positions. If someone, a leader or otherwise, isn't as compassionate as some might like or as humble as others might like, it is not uncommon for that person's character to be wholly condemned. Is that reasonable?

When we think about the virtues that compose character, it is clear that while they are interrelated, they also can exist independently. Since virtues can be independent, that means they cannot be all-or-nothing; it is possible that one virtue can exist while another may be absent. The same follows for character, too; it must not be viewed as an all-or-nothing concept. Someone can have great personal integrity but be woefully lacking in courage. Most of us know someone who is

★ ☆ ☆ ☆ ★

largely irresponsible or lacking dependability yet is supremely compassionate or honest. Could such people be considered to have some degree of high moral character? Or, should we judge someone as wholly immoral because they habitually fail to keep appointments or don't express their emotions the way we like? General Glenn Walters, former Assistant Commandant of the U.S. Marine Corps and President of The Citadel, said what we all know: "no human being [...] is perfect. You're going to screw it up." Recognizing this, we cannot expect either ourselves or others to possess and/or display all of the virtues of character 100% of the time. That doesn't mean we shouldn't strive to do so, especially as leaders. It just means we are going to fall short now and again.

Let's return to the idea of character existing on a spectrum because it is vital to understand this if we are to effectively lead others. In their 1998 study of children and adolescents, Clark Power and Vladimir Khmelkov found that character development follows cognitive development. Additionally, in a study of youngsters aged 8-to-16 years, Angela Evans and Kang Lee showed that intentional practice of a virtue helps increase that virtue, such as when children promise to tell the truth, they are 8-times more likely to be honest. These studies show that character and the virtues that comprise it are manifested or mastered along a spectrum. That fits with our everyday experience. Since we know moral perfection is not humanly attainable, and we also know there are both people who have high and low moral character, it makes sense that it occurs on a spectrum.

Let's imagine that spectrum is a scale from 1 to 10, with 1 being the lowest possible manifestation of a given virtue, and 10 being the highest. Someone could be highly courageous, scoring a 10 on our scale, and yet be minimally trustworthy, scoring only a 2. Overall, that person may well not be of high moral character, even though they scored a 10 in courage. An example might be a criminal who has the courage to rob a bank but has few other virtues. We wouldn't consider that person of high moral character. Conversely, another person could have courage and trustworthiness that fall around a 7 on the scale—not topping out the scale, but consistently above average. In that scenario, it is

★

possible we would consider them to have fairly high moral character. So, when we think of someone with high moral character, we are typically envisioning a person who consistently manifests the virtues on the high-end of the spectrum, though some, if not all, are manifested imperfectly. This idea of the spectrum of character and virtues is particularly important for us as leaders; while we are allowed minor mistakes, it may only take one area of major character weakness to rob us of the credibility we need to lead well.

The good news is character and the virtues that comprise it can be developed and/or improved, which was something multiple Four-Stars touched upon. For instance, General George Casey, 36th Chief of Staff of the U.S. Army, explained character development is a practice, saying, "You build your character over time by habitually trying to do the right thing." This idea that character and its components are developed is important not only for the leader, but also for the leader's understanding of the led. General Glenn Walters discussed this with regard to how he develops the perspectives of the students who serve on The Citadel's Honor Court. He said, as with most universities, The Citadel has some students "who were raised very, very well," as well as some students who "might not have had the right role models, but they're smart, so we're going to give them a chance." For those students who have not had character modeled and taught to them, General Walters insisted that the student body, faculty, and university leadership "have to fill that void" by being role models of character, "because we understand that not everybody has the same basis" for understanding what character is. As evidenced by the long tradition of graduates who have gone onto leadership roles, The Citadel has been successful at enhancing the character of its students and faculty. It does so by expecting high moral character from all its members, and ensuring they hold each other accountable to consistently model honor, integrity, duty, loyalty, trustworthiness, and the other virtues of character. This commitment is crystallized in The Citadel's dedication to the Honor Code: "I will not lie, cheat, or steal, nor tolerate those who do."

Beyond the influence of role models and practicing the virtues of character, if we are to learn, develop, and improve our character, we

★ ☆ ☆ ☆

must be able to recognize the areas in which we need to improve. That recognition requires our identification of a personal deficit—if you need to learn calculus, you are, by definition, deficient in calculus. To be able to identify and admit a personal deficit requires that we know our strengths and weaknesses; it is therefore imperative that we possess self-knowledge, as well as the ability to be honest with ourselves.

HOW WELL DO YOU KNOW YOURSELF AND YOUR CHARACTER?

Inscribed on the wall in the court of the Temple at Delphi, and wrestled with by Socrates, the admonition to "know thyself" is millennia old and is found across multiple ancient cultures. Within the leadership literature, as far back as 2,400 years ago, Sun Tzu admonished his reader, "Know thy enemy and know thyself, and you should not fear the outcome in a hundred battles." The idea is both incredibly simple and terribly difficult. If you, as a leader, know your strengths and weaknesses, and the strengths and weaknesses of your opponent, then you can capitalize on your strengths; guard against your weaknesses; capitalize on your opponent's weakness; and guard against your opponent's strengths. In so doing, according to Sun Tzu, you will be successful. That seems simple and makes sense. You can study your opponent and learn their ways, their observable strengths and weaknesses. But it is far more challenging to objectively study and know yourself.

Most of the Four-Stars discussed the importance of knowing yourself as fundamental to strong personal character. In fact, nearly two-thirds of them discussed the vital role it plays in successful leadership. Many of them recounted stories from their careers where knowing their own strengths and weaknesses was crucial. For instance, when asked about a leadership lesson he learned the hard way, General Pete Pace spoke specifically about his failure to "understand who [he was] as a leader and as an individual," and how that failure nearly resulted in him committing an atrocity.

★ ☆ ☆ ☆

General Pete Pace, *on Knowing Yourself and Preventing Character Failure*

General Pete Pace was the 16th Chairman of the Joint Chiefs of Staff. But during a leadership class at Marine Basic Officer School, then-Lieutenant Pete Pace learned that sometimes in combat, troops in the past had committed atrocities, including killing innocent women and children. As he sat through the class, he thought to himself, "How can that be? I would never, ever allow myself or my Marines to do anything illegal or immoral in combat." With that conviction still ringing in his mind, Lieutenant Pace shipped off to Vietnam to take part in the Tet Offensive of 1968.

On July 30, 1968, Lt. Pace was leading his platoon across a clearing toward a village outside of Da Nang when shots rang out.

> An enemy sniper shoots my machine gun squad leader right smack in the chest from a village. [He was a] good young man named Lance Corporal Guido Farinaro. Guido was 19-years-old. I got to Guido before he died, and when he died, I was furious. So, I called in an artillery strike on the village. Fortunately, [...] it takes a while for the guns to understand where you are [and] where the enemy is. [They] had to put the right elevation [...] on the guns. That took some time. During that time, my platoon sergeant, my senior enlisted Marine didn't say anything to me, but I could tell by the way he was looking at me that I was doing the wrong thing. So fortunately, I was able to call off the strike and do what I should have done in the first place, which was sweep through the village on foot. [...] all we found was women and children. The sniper was long gone, long gone.
>
> I don't know how I could be living with myself today had I slaughtered those innocents. It stunned me that I'd almost done that. When we got done with the patrol, I got my platoon together, and I apologized to them. I told them, as best I could, I would never let that happen again.

When he left for Vietnam, Lieutenant Pace thought he knew himself; he was certain his character would never allow him to do something so horrible as killing innocent people. Yet, in the heat of the moment his weakness was revealed, and it nearly cost the lives of a village full of women and children. Following that watershed moment, General Pace has made it an ongoing practice to set up mechanisms to protect him from ever doing something like that again.

★☆☆☆

Every day since 1968, to include this morning, while shaving, I think about what might challenge me morally today. 99.999% of the time, nothing challenges me morally today, but that's not the point. The point is, just like in any education, you train yourself how to think. I've trained my brain, if I'm thinking about giving an order, to say, "Is this legal? Is it moral?" If I'm receiving an order, "Is it legal, and is it moral?" It takes three nanoseconds to come to that conclusion. It's really a very quick process, but you've got to have it in your head.

General Pace also emphasized the importance of this sort of mental process beyond combat.

It's not only about life and death situations like I faced. [...] In everyday life, even in corporate life today, most people don't come to work saying, "I think I'll do something illegal. I think I'll do something immoral." No. What happens is good people come to work, and somebody rolls a sweat grenade in the room, and something happens. A competitor does something, an enemy does something. All of a sudden, you've got to react. You've got to react quickly. That's when having the anchor of a moral compass is really important, because you start cutting corners without even realizing you're cutting corners.

To help maintain his moral bearing and character, General Pace practiced mutual accountability with those he led and recommended the same for others.

When you take command, tell the troops, "I am going to lead as best I can in a moral and ethical manner. I expect you to do the same. If I see something you're doing that I think is not right, I'm going to call you on it, and I need you to call me on it." I'm challenged most morally when I'm least prepared emotionally to deal with it. My Marine gets shot; I'm furious. Something happens that needs to be taken care of right away, I'm moving too fast. Those are the things that happen to people, and they're not making conscious decisions to do it wrong. It's just happening, because they haven't set up their left and right limits [...] and asked their subordinates to help them stay on that path.

DOES CHARACTER MATTER IN LEADERSHIP?

Having identified what character and its constituents are, as well as how those can be self- and others-related, we reach the question of

"Does it really matter?" Specifically, is high moral character an immutable contributor to excellent leadership? Further, if it is a non-negotiable component, why is that the case?

In the Introduction, I made the case that, apart from esoteric discussions in academic leadership circles, most people do not consider leadership as a value-neutral thing. Instead, they think of leadership as being about accomplishing something worthwhile or good. Also, it is usually the case that someone is judged as displaying bad leadership when they don't have or have failed to demonstrate various virtues, or their actions have caused the deterioration of their group or organization. If it is true that the concept of leadership carries intrinsic positive moral value, then it follows that to be a good leader the person carrying out the leadership must have and demonstrate some positive moral constituency; that is, character.

You may be familiar with Peter Drucker's statement that "Managers do things right, while leaders do the right thing." This is a truism that communicates the relationship between leadership and character. From the perspective of the Four-Stars, leadership and character aren't merely related—most of them view the two as inseparable. Beyond the inseparability, two perspectives emerged from the Four-Stars that solidified the central role character plays in leadership. First, some insisted that it is not the outcome of your leadership that matters as much as how you achieve those outcomes. General Darren McDew, former Commander of U.S. Transportation Command, framed this belief concisely by saying, "The issue is how, not what you get done," and General Marty Dempsey, the 18th Chairman of the Joint Chiefs of Staff, mirrored that when he said, "It's not just what we accomplish, but how." Most people know the value of character on some level. For instance, when a traveling carnival comes to town with their menagerie of rides, games, and oddities, that organization isn't typically viewed as an exemplar of business ethics. They may rake in a lot of money, but we know they're dubious, especially in the games they offer. The basketball goals are too small and too high. The metal bottles are magnetized. The darts are dull, and the balloons seem to be made of Kevlar. Though those carnivals may be making money, they aren't doing it in

★ ☆ ☆ ☆

the right way—they're cheaters—and most people don't like that. It's the same with leaders—if you are immoral or unethical, though you may achieve some desired outcome, most people won't follow you.

For the Four-Stars, the high moral standard for leadership didn't stop there. Others extended it further to include not just accomplishing the right objective in a moral and ethical manner, but also doing so for the right reason, free of ulterior motives. General Ann Dunwoody, the first female Four-Star officer in the U.S. Military, verbalized that clearly when she said a leader of character will "do the right thing for the right reasons." Similarly, General George Casey explained that one's character is built over time through "the habit of doing the right things for the right reasons." As an example, think about a young man who takes his date out to an expensive dinner and concert. So far, he may have done the right thing (i.e. treated her nicely and spent time with her) in the right way (having an enjoyable dinner and attending a concert). If he did those things for the simple reason that he enjoys spending time with her and wanted to do something special with her, we would say he did it for the right reasons and see that as good. But, if his whole purpose was to coerce her into sex, we would think he has no character, even though he did the right thing in the right way. In the same way, leadership of the highest caliber doesn't just accomplish the right thing—the mission. Further, it doesn't only accomplish the right thing the right way. Rather, the zenith of leadership is to accomplish the right thing, the right way, for the right reason(s). Think back to General Krulak's example of then-Brigadier General Jim Mattis. General Mattis did the right thing (gave Christmas off to the captain with kids) in the right way (General Mattis took the duty himself without making a show of it) for the right reason (because General Mattis wanted the captain be able to spend the holiday with his kids). Figure 1.2 displays a visual demonstration of this concept and how operating with the highest character allows you to achieve excellent leadership.

★ ☆ ☆ ☆

Figure 1.2. The Influence of Character on Achieving the Highest Levels of Leadership.
Good leadership is achieved when a leader does the right thing. Great leadership is
reached when a leader's character pushes them to do the right thing for the right
reason(s). Four-Star leadership is only attained when a leader's character drives them to do
the right thing in the right way for the right reason(s).

The Four-Stars' answers to the third question in my interview revealed
their perspective on how important character is for leadership. I asked
each of them how they want to be remembered as a leader—what
would they want their leadership legacy to be? When faced with that
question, most paused reflectively. Over half responded by discussing
character and the virtues that comprise it in what they hope for their
legacy to be. For instance, General David Rodriguez, former Commander of U.S. Africa Command, wanted his leadership legacy to include
being a person with "high morals, high integrity, and good character," and General J.D. Thurman, former Commander of United Nations
Command, Republic of Korea-U.S. Combined Forces Command, and
U.S. Forces Korea, wanted to be remembered as a leader who, among
other things, "was always ethically and morally straight." Among all the
Four-Stars, the only thing more commonly desired than character for

defining their leadership legacies was to be remembered for caring about those they led. For the Four-Stars, it was resoundingly clear that high moral character not only matters, but for many of them, it is the most important determinant of excellent leadership.

WHY DOES CHARACTER MATTER?

Recognizing that character is important for leadership, the next step is to understand why that is the case. What does character contribute to excellent leadership?

Fundamentally, leadership is about a relationship between you as the leader and those you are leading. The nature and strength of that relationship allows you to have influence on those you lead in such a way that you can achieve a common goal, which is the functional definition of leadership. Our relationships, be they with family, friends, or coworkers, are built on things like integrity, honesty, trustworthiness, and mutual respect—the kind of virtues that comprise character. Without those, there is duplicity, distrust, and disrespect, which destroy relationships. So to have effective relationships of any type, including those that allow you to influence those you lead, you must have character and some of its constituent virtues.

For the Four-Stars, the quality of a leader's character is crucial to the relationship of those they lead. Some described the positive impacts that character has on both the people you lead and the accomplishment of the mission. General Stanley McChrystal, former Commander of Joint Special Operations Command who oversaw the capture of Saddam Hussein, explained that a leader with strong values of character, including integrity, consistency, and kindness, "becomes much more effective and much more able to do things." General George Casey said, "People trust men and women of strong character to do the right thing in difficult times—to do the right thing for the organization and not themselves." He described the trust between the leader of character and the led as "the glue that binds the whole organization together." He concluded saying, "Leaders with strong values build strong organizations, period."

★ ☆ ☆ ☆

The Four-Stars also described the negative impacts of a leader's absence of character. As he talked about the role of character and trust in successful leadership, General Chuck Wald, who served as the Deputy Commander of Headquarters U.S. European Command, said, "You've got to be trusted and believed. If [character and integrity] aren't there [...] people won't trust you, and they won't follow you even though they think you have a lot of good things going for you." General Wald went on to say that if people don't trust their leader, "They're probably going to fail you in the critical moment." This point of failure at the critical moment was also raised by General Robert Kehler. He explained that it is essential for leaders to have character so that those they lead will have trust and confidence in their leader, because if they don't, "at a critical moment, someone will question what they're being asked to do, or they'll refuse to do it." Admiral James Hogg, who commanded the U.S. Navy's Seventh Fleet, was more explicit about the role of a leader's character, the call of duty on the led, and the negative impacts of the absence of character, saying, "If you don't have [character], it's difficult for you to lead your troops, especially into combat." He explained that troops will recognize, sometimes unconsciously, an absence of character in their leader and have distrust, even though they may not realize why they do not trust the leader. As a result, Admiral Hogg said, "It is a problem for them when the time comes to put [themselves] in the position where [they] might be making the ultimate sacrifice."

The Four-Stars made it clear that character is vital for excellent leadership. It nurtures the trust and confidence that are foundational for the relationships on which effective leadership is based. The presence of high moral character in the leader creates strong and effective groups and organizations, whereas its absence leads to dysfunctional groups prone to failure at crucial moments. This is because no one wants to follow someone they do not trust, especially when following is a matter of life and death. The Four-Stars' perspective on the importance of character in leadership was consistent with Edgar Puryear's conclusion in his 1992 book *19 Stars*: "...you can have character without leadership, but not leadership without character."

★☆☆☆

General Charles Krulak, *on Character in Difficult Circumstances*

Just two months after becoming the Commandant of the United States Marine Corps, General Charles Krulak found himself navigating an extraordinarily charged and delicate leadership situation with high-stakes, international consequences. On September 5, 1995, two U.S. Marines and a Navy Corpsman on Okinawa kidnapped and raped a 12-year-old Japanese girl, and outrage gripped Okinawa and Japan. The Japanese government was furious over the incident, and the U.S. Congress was concerned about the impact on U.S.-Japanese relations. With various members of the U.S. government trying to determine how to address the situation, General Krulak felt personally compelled to go to Okinawa. He requested that the Secretary of the Navy allow him to go, but was told, "There's no way I'm sending you out there. It is so volatile. We're going to cause all kinds of problems." General Krulak then requested to meet with the Secretary of Defense, Bill Perry, who told him to "Go out there."

When he arrived on Okinawa, General Krulak recalled, "I talked to every single Marine and sailor that I was in charge of on that island in a 24-hour period—every one of them." He told his Marines, "I believe in you. We had a terrible thing; it was not you. [...] Your America and your Marines believe in you and believe in what you're doing." Following his meetings with all the Marines on Okinawa, General Krulak went to meet the governor of the Okinawan prefecture. He had been told by numerous people, "You can't go see the governor, or he is going to just use this as a media ploy." However, Krulak felt strongly compelled to meet with him anyway.

When he entered the prefecture office, General Krulak found the governor had packed the hallway with members of the media. Making his way through the crowd of people in the hall, General Krulak reached the governor who "stuck out his hand to shake. I took his hand and pulled him towards me and hugged him, and the picture that was all over the paper was the Governor and the Commandant hugging each other." Subsequently, the Chairman of the Japanese Self-Defense Force came to the Pentagon, and General Krulak asked to meet with

★

him. General Krulak went to his office and donned his full dress-blues, medals, and sword. When the Japanese Chairman walked in, General Krulak bowed and said, "I want you to understand the deep sadness that the Marine Corps has that this happened. I apologize, but it is not my Marines. It was two [Marines] and one [Navy corpsman]. We will punish them, and you will punish them. But please understand and accept our apologies."

What was it that compelled General Krulak to step into the leadership breech between the U.S. and Japan? Or to go in-person to Okinawa to encourage his Marines in the midst of the volatile public outcry against them, and to humble himself before the Japanese officials and ask for forgiveness? "It was just a sense [...] that character, at the end of the day, is going to win out over anything." His sentiment echoes what the U.S. Army Leadership Field Manual says: "People of integrity do the right thing not because it's convenient or because they have no choice. They choose the right thing because their character permits no less."

Having addressed the fundamental nature of character and its importance in Four-Star leadership, let's now turn to what it looks like to be a leader of character.

★ ★ ★ ★

CHAPTER 2

BE THE EXAMPLE

*Always lead by example. So, [that means] the way you treat others,
the way you speak, the way you interact. You want to always project the
values, the style, the approach that you want from others.*

– ADMIRAL MICHAEL ROGERS, U.S. NAVY –

A s Commander of both Joint Task Force-Southwest Asia and the 9th Air and Space Expeditionary Task Force-Southwest Asia, then-Major General Gene Renuart had a lot of responsibilities and demands on his time. At the same time, being an example to those he led was something he saw as fundamental to being a leader. As a pilot, that meant taking his turn flying combat missions over Iraq. He did so recognizing that if he were killed or captured, it would have much larger consequences for the mission and Air Force than for lower ranking officers.

> As a Two-Star, I flew combat missions in the F-16, much to the chagrin of some of my bosses. But okay, if I'm going to send captains and lieutenants out there and I'm qualified in the airplane, why should I not be out there leading? I didn't do it every day. I wasn't reckless. I understood the value of having someone in that role shot down or captured [...] but I felt it really important that the people I was sending out every day to conduct those missions saw me going out to do that.

★ ☆ ☆ ☆

Being the example meant more to General Renuart than just being out flying the missions; being the example meant doing it well.

> I took the same evaluation test. I was a flight examiner. I was a four-ship flight lead. I did all the same things as my experienced captains, majors, and lieutenant colonels did, because I felt it was important for the lieutenants to see the old man, if you will, struggling through the daily bold face test or the closed book exams, or the annual instrument test, or all those things, and being able to defend tactics in the briefing room.

Because he flew combat missions as a commander, the junior pilots respected him as a pilot, which fostered trust and confidence in him. That trust and confidence helped General Renuart build camaraderie with his officers, and it also gave him the opportunity to instill in them a greater understanding of what excellent leadership looks like.

> It's that kind of leadership that I think leaves a lasting impression on individuals that will serve them well for their decades to come and hopefully, [they will] embrace some of those same sorts of things.

If we are to be great leaders, we must be the example for those we lead. This idea is so essential to great leadership that nearly three-quarters of Four-Stars specifically addressed it. In fact, it is so important to leadership that the entire content of all the interviews with Four-Stars, and subsequently the content of this book, could be nested within it—a leader must be the example. While it is probably daunting to consider, the reality remains: if you are the leader, you set the standard for those you lead. Over and over, the Four-Stars insisted, if you strive for excellence, so will those you lead. If you tolerate sloppy work, so will they. What you do becomes acceptable practice, whether it is building people up or tearing them down. Your people look to you to determine how they should act—how they should be—within the group or organization. There may be no greater refinery for character than to be expected to serve as the example for others. When you take on the role of being a leader, you take on that responsibility, whether you want to or not. You are the example.

A part of being the example is understanding that the people you lead have an expectation of what a leader should be. General Tony Zinni said the people you lead "have an image of you that sometimes could

★ ☆ ☆ ☆

be a tough one to live up to." He emphasized that being the example means always living up to that responsibility, because "even when you don't think people are watching you or paying attention to you, they are." General Charles Krulak said the same thing. When I asked him to recall five leadership maxims he always tries to keep in mind, his first response was, "Your people are always looking at you." He followed, "People are looking at you all the time—seniors, contemporaries, and juniors are looking at you all the time. If you walk into a room with a lack of energy, lack of optimism, not understanding authority and responsibility and accountability," it has a negative impact on your people and decreases your ability to lead. As a leader, you must, at all times, be the example to your people that you want them to see.

The Four-Stars consistently identified three major ways leaders must be the example for their people: do the right thing; do your best; and never walk by a problem. Figure 2.1 shows that these three principles intersect in *being the example*. These, along with supporting concepts, will be covered sequentially in the balance of this chapter.

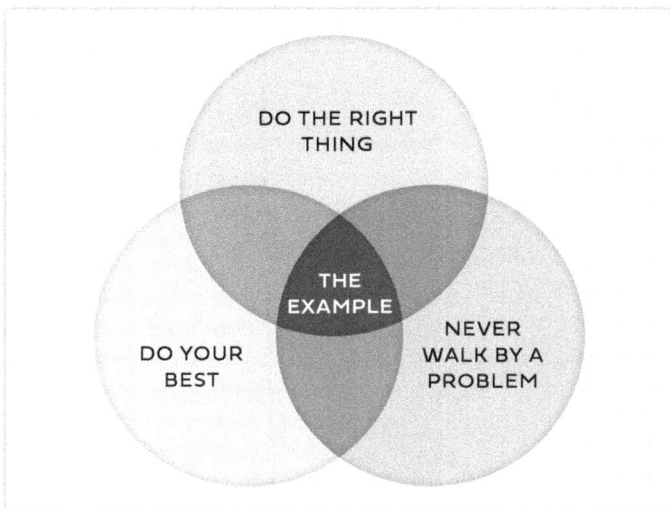

Figure 2.1. The Components of Being the Example. Being the example requires doing the right thing, doing your best, and never walking by a problem. Doing any of the three in isolation can be helpful, but also comes with risk. For instance, if your focus is only on doing your best, you may be viewed as a self-focused achiever. If your focus is only on doing the right thing, you may be seen as prudish. If you fail to do the right thing and/or your best, but you never walk by a problem, you are at risk for being seen as judgmental.

★

WHAT DOES IT MEAN TO DO THE RIGHT THING?

In 1983, then-Major Tom Hill was Aide-de-Camp to the Chief of Staff of the Army, General John Wickham, serving in the Pentagon. When he came into his position, Major Hill found in his office a large, standup wooden desk, which had been the standard issue for people throughout the Pentagon for decades. Shortly after Major Hill moved into his office, General Wickham shared a story about those desks with Major Hill from when he had been the executive officer for General Harold K. Johnson during the Vietnam War. As General Hill recounted, "General Wickham said he'd walk in at nine, 10 o'clock at night sometimes, and he'd find General Johnson standing there [at his wooden desk] writing these notes [to families of soldiers killed in Vietnam], crying." It was a powerful image of the weight of leadership and the character of General Johnson. Major Hill understood why General Wickham revered General Johnson so much, and it fostered in him the same respect and admiration for General Johnson.

Sometime later, General Johnson's health began deteriorating, and he was hospitalized at Walter Reed Army Hospital. General Wickham and Major Hill were set to travel to Europe when Major Hill received an urgent phone call from the Surgeon General of the Army.

> "Tom, you need to get General Wickham over here." He knew that Wickham and Harold K. Johnson were close. He said, "You need to get him over here to see General Johnson, because he will not live longer than the trip you're going on or very shortly after that." So, we went over there, and there's three people in the room. It's Harold K. Johnson dying, General Wickham sitting next to him in a chair, and me off in a corner. General Johnson looked at General Wickham and said, "There's only one thing I regret in life." He said, "Earle Wheeler," who was the Chairman of the Joint Chiefs of Staff, "called the meeting of all the Chiefs on a Saturday morning at Fort Myer in the Johnson era. [We] all decided that [we] were so sick of what was going on in Vietnam that on Monday morning [we] would go over to the White House and hand the President of the United States all of [our] stars—quit *en masse*. On Sunday, General Wheeler called the Chiefs and said, 'No, we're not doing this. One, it's tantamount to mutiny, and we can't do that. That's not our oath. Secondly,' he said, 'all he would do is replace us with the people who are going to do his bidding anyway.

★ ☆ ☆ ☆

We need to continue to do this as best we can.'" Then, [General Johnson] looked at General Wickham, and he said, "My only regret is we should have done that. Think of all the lives we might have saved."

As with Generals Wheeler and Johnson, leaders often face moral and ethical dilemmas that can be wickedly complex, and it can be hard to determine what it means to do the right thing. General Wheeler determined, from his perspective, the right thing to do was to remain in their roles because at least they would have the opportunity to make some positive impact on the outcome, whereas General Johnson carried a deep regret that he did not hand over his stars to President Lyndon Johnson. But, what is the "right thing?" How can a leader know what is right?

For millennia, societies have codified rules of behavior—what it looks like to do right by God, others, and self. Certainly, there are tomes and scriptures expounding on the question, all of which far surpass what can or should be covered here. However, there are a couple of elementary, bedrock principles that can be summarized in addressing the question. First, we should have a standard that is external to ourselves. Second, we are each a part of a larger society and cannot operate without regard to that.

These two principles were summarized by Jesus Christ when He was asked what the most important law was. He summed up the entirety of the law in two statements: "Love the Lord your God with all your heart, soul, mind, and strength," and "Love your neighbor as yourself." (Gospel of Mark 12:30-31) Recognizing there are many who do not believe in a higher power or religious texts, His answer addresses both principles concisely. He acknowledged a need to recognize and adhere to an external fixed standard, while also acknowledging we cannot operate in isolation from others.

Why should we need to determine goodness or rightness based not on ourselves but rather some external standard? Because we, as humans, carry a host of cognitive biases, many of which function nearly exclusively to protect our egos (e.g. confirmation bias, Dunning-Kruger effect, etc). Our subconscious does all it can to make us think we are right, especially when we are wrong. Apart from an external

★ ★ ★ ★

standard, these cognitive dispositions cloud our ability to know what is right, because they drive us to think what we want—what makes our egos happy—is right. An external, fixed standard provides something to weigh our assessment against; a way, if we are willing, to be honest with ourselves. This external, fixed standard does not change, unlike our whims, opinions, and personal preferences. Instead, it provides us with a stable means of knowing what is right.

The second principle we see in Jesus' answer is, as a part of society, we do not—cannot—operate in isolation. Being right means recognizing the inherent worth of others as equal to our own and parsing decisions through that lens. It means treating others with dignity and respect, a principle consistently reiterated by the Four-Stars, and considering whether our proposed actions will do so. Doing the right thing seldom means doing that which pleases or advances you at someone else's expense.

Reflecting on "doing the right thing" and the moral complexity that can come with it, General Hill paused, reached into his wallet and pulled out a now-laminated piece of paper, yellowed and creased with wear from years of repeatedly taking it out to re-read what he had written on it in 1968 when he was as an infantry officer in Vietnam. Once more, he read the quote aloud.

> "Cowardice asks the question, 'Is it safe?' Expediency asks the question, 'Is it politic?' Vanity asks the question, 'Is it popular?' But conscience asks the question, 'Is it right?' And there comes a time when one must take a position that is neither safe, nor politic, nor popular, but he must take it because conscience tells him that it is right."   – Martin Luther King, Jr. –

Our conscience tells us what is right, and our character empowers us to be able to do the right thing. The two are entwined—one informs the other. Without integrity of character, we will not do what our conscience tells us is right. Among the nearly two-thirds of Four-Stars who discussed doing the right thing as fundamental to being the example, essentially all framed it in the context of character. General Mike Scaparrotti, formerly the Supreme Allied Commander, Europe (SACEUR), exemplified this relationship when he said, "Character's a hard thing

★☆☆☆☆

to define, but it's broader than just integrity of word and deed. It has to do with this innate sense of doing what's 'right.' That isn't just a legal term. It isn't just an ethical term. It's the sense of doing right—morally [and] ethically, as well. To me, the first signal of that is whether you lead by example." In line with General Scaparrotti's identification of the relationship among character, leading by example, and doing the right thing, the Four-Stars described multiple components of character that contribute to one's ability to do the right thing consistently. Among those virtues were integrity, honesty, consistency of action, and dependability.

As the Four-Stars discussed the virtues of character needed to be the example, it reinforced the idea from the prior chapter that they are interdependent—they flow from one into the other. This concept of the interconnectedness of virtues isn't new—it dates back to Aristotle and Plato, and was known as the Unity of Virtues. For the ancient Greeks and Stoics, virtue was a singular whole; a person could not have one virtue without the others. For instance, someone couldn't have honor without morality, as that might lead to vengeance. Similarly, courage without wisdom would likely produce foolishness. In this way, all of the virtues are integrated into a single idea, which, within the auspices of Four-Star leadership, is character.

Since character represents the composite of integrated virtues, it stands to reason that integrity was commonly discussed by the Four-Stars and was seen as imperative. Reflecting on integrity, General Charles Krulak said, "Integrity: It's the only thing you own. [...] No one can take it from you, but you can give it away," and, he said, "when you give it away, it is really hard to get back." Integrity is the irreducible component of character; without it, a leader will not remain honest, dependable, consistent in word and deed, fair, selfless, or trustworthy. These all derive from integrity, which General Ed Rice said is "about doing the right thing." In addition to those virtues of character, over three-quarters of the Four-Stars included treating others with dignity and respect as being part of excellent leadership and doing the right thing.

★ ☆ ☆ ☆

WHAT DOES IT MEAN TO DO YOUR BEST?

After 32 years of service, General Mike Scaparrotti's father retired from the U.S. Army as an E-8 company first sergeant, a rank lower than is typical for someone with such a long service career. It turned out that it was because he wanted to do his best in his job, which meant foregoing being promoted in exchange for doing what he loved.

> I tend to believe that you have God-given talents, and there are limits to that. So, my father [...] realized, "I'm really good at being a company first sergeant. That's what I love. I don't really want to be a sergeant major, because I'm good at this." He could have been a sergeant major, but he knew that he wouldn't be as good as he [was] right there, and that's all he did, and he left the military as an E-8. I respect that. I think everybody ought to try and understand, individually, if they're a leader that cares about their commitment to their subordinates, they ought to really seek that out.

Recognizing his father's commitment to doing his best led General Scaparrotti to begin to seek out what it would take for him to do the same—his best. As he moved through his career, General Scaparrotti began to frame the idea of doing his best from the perspective of achieving his personal potential. He recognized that to achieve his potential—to do his best—he would have to push against his self-imposed limits.

> The way I put this to people is, I've tried to find the limits of my potential. I tell them, "Try to find the limits of your potential and understand what they are. It's probably not nearly as close as you think. You're capable of a whole lot more than you think, but you also have to be willing to be candid with yourself and change."

As General Scaparrotti was trying to delimit his potential, he realized that he would never reach it without making changes.

> There is no way that I would've made it even to Two-Star [general] or beyond had I not been willing to change some of my habits and learn. Because, [...] there is a big difference between colonel and brigadier general [despite colonel being only one rank below brigadier general]. Then, my experience was there is a difference between One-Star (brigadier general), Two-Star (major general), Three-Star (lieutenant general), and

★ ☆ ☆ ☆

Four-Star (general), particularly if you're leading deployments like we just happened to be doing over the entire time that I was a General officer. So yes, there is a limit to potential, [...] but if they're not willing to work and change, they're not going to reach that potential, and they can't serve in those positions in the way that we would like them to serve. [...] Unfortunately, far too many people don't really understand their potential.

If we are to be the best possible leaders we can be—if we are to reach our full potential—we have to be willing to change, to work hard to improve our capabilities, and jettison those things that hold us back. General Scaparrotti realized to be able to do his best meant working to achieve his maximum potential, which meant making necessary, though sometimes difficult, changes. Many leaders continue to rely solely on the capabilities that got them to where they are, never reaching their highest leadership potential. This was recognized by Marshall Goldsmith and addressed in his popular book, *What Got You Here, Won't Get You There*. Great leaders are able to recognize and admit their need to change so that they can realize their full potential and truly do their best.

What does doing our best really look like? Does doing your best mean operating at 100% of your physical, mental, and emotional capability all the time? Obviously, no. In fact, it is rare to operate simultaneously at 100% in all three; this perhaps only occurs when we enter a state of what Mihaly Csikszentmihalyi describes as "flow." So, aiming for 100% all the time is unrealistic, even foolhardy. Instead, to do your best more accurately means doing the very best you can in any given moment when you consider the physical ailments, mental distractions, and emotional weight you are experiencing. As an example, imagine a sprinter who can run 100 meters in 10.3 seconds. Now, imagine we encumber her with a 50-pound weight vest. If she runs with all her might, giving everything she's got, she will not be able to run the 100 meters in 10.3 seconds. That does not mean she didn't do her best; it means she did her best given the circumstances. In discussing doing your best regardless of the circumstance, General Pete Chiarelli said, "No matter what job you're put in, whether it's the one you wanted or the one you were dreading getting, give it your best. Figure out a way to do it better than anybody's done it before, even though it might not be the thing you wanted to do. I think that's absolutely critical, that you always do your best no matter what the position that you find your-self in." Those people who do their best are viewed more favorably, get more opportunities, and are generally more successful.

The difficulty comes when we start using some circumstance—an ailment, distraction, etc.—as an excuse for not doing our best. General Ed Rice said, "I think leaders really have to be the person in an or-ganization that doesn't allow the organization to find excuses for not getting something done." How can we avoid allowing ourselves the ex-cuse? I identified three major practices that help the Four-Stars avoid excusing themselves from doing their best: self-discipline, owning their mistakes, and asking for help.

Don't abandon self-discipline

As a brand-new platoon leader, then-Lieutenant Stanley McChrystal faced a dilemma of self-discipline and doing the right thing during a training exercise. The purpose of the exercise was to arrive at a speci-fied destination, set up their mortar batteries, and then fire on targets downrange.

> I took this mortar platoon in the 82nd Airborne, this is 1970s paratroopers. It wasn't the great Army then. We went out, we go to this position, and we dig our mortar pits in, which takes several hours. It's hard work, and it's raining. So, we do this, and we get all dug in because I'm adhering to standards. [...] Then, we [contacted] range control which allows you to live-fire, which is what we're out for, and they said, "You're at the wrong firing position. It was coordinated incorrectly, so you've got to move." So, now we're about two in the morning. We had to move the platoon, and now we go to this new position, and the question is, "Do we dig in again?" Digging in's the right thing to do, [...] but now everybody's exhausted, and the screw up was mine, the scheduling thing. So, we had to move because of me. [...] So, that was one of those moral questions. I had them dig in. I was not popular in the moment, and I dug with them. But when I look back [...] that was the only right answer. But there's little chances like that almost every day through a career. [...] it's the self-discipline to stick with what you think your values are, what you believe [is right].

★ ☆ ☆ ☆

Students of stoicism and other ancient philosophies understand that self-discipline is grounded in knowing what you're capable of—knowing yourself—and then making it an ongoing commitment and practice of working to achieve it. When we know what we are capable of, self-discipline is the drive—the commitment—to not give up on ourselves, to not sell ourselves short, and to do the right thing. It is the commitment to do what it takes—the hard things—over and over so we can become what we are capable of being.

As leaders, self-discipline is also the ability to extend that commitment to those we lead. It is to support them in becoming what they are capable of being, too. As a leader, it can be tempting and easy to let people settle for good enough, especially when they are giving you a lot of negative feedback (a.k.a. complaining). For the leader, self-discipline is not only pursuing our personal best, but also staying true to the role of leadership, what General Gus Perna repeatedly phrased as "being responsible and accountable to the role." Undoubtedly, plenty of leaders fall short because they do not have the self-discipline to keep their people moving forward when they start meeting resistance.

Own your mistakes

Returning from Vietnam, then-Major Barry McCaffrey thought he would have some time to recuperate. Instead, he was assigned a larger duty that revealed some of his weaknesses.

> As a brand-new major, I was suddenly sent back to take command of a battalion. So suddenly, I was thrust into this immense responsibility. They immediately deployed [to] Hohenfels [for] a graded exercise, day [and] night [for] five days. On the fourth day, [I was] totally exhausted, incoherent, [and] having a ball. I loved it. I was running the final attack, and the attack was totally hosed up. It was so screwed up, it was impossible to repair it. [My division commander] pulls up next to me, and he knew the thing was completely hooked up. He said, "Well, Barry, how's it going?" I remember saying, "Well, sir, I'm running the most [fouled]-up attack ever seen at Hohenfels Training Area, and I'm trying to get it under control, so I can't talk to you any longer," and he drove off.

★

Excellent leaders have the strength of character to admit their mistakes. Over and over, the Four-Stars discussed the necessity for leaders to own their mistakes. There are two important reasons for this. First, recognizing and owning our mistakes, failures, and shortcomings is the gateway to learning, growing, and improving. Fundamentally, if you do everything right and always have the right answer—if you never make a mistake—you will never learn. When we are learning to walk, we fall down; we get back up; we fall down; we get back up, over and over, each time learning more about what it takes to walk. If we fall down and do not recognize it (own it), remaining down, we will never learn to walk, let alone run. This extends to any endeavor, physical or cognitive. If we cannot see and admit where we have fallen short, we cannot learn from it and continue to grow toward what we are capable of being—our best.

Further, owning our mistakes demonstrates at least two important things to those we lead. It shows them that admitting mistakes and learning are important, and it demonstrates we are trustworthy because doing the right thing—owning our mistakes—is more important to us than thinking we are right. Many leaders try to maintain an image of infallibility for fear no one will want to follow them otherwise. That outlook is misguided. People know you are not infallible, and we don't trust people who act as though they are. Without trust, it is extraordinarily difficult to lead successfully. Perhaps revealing the power of owning our mistakes, when President John F. Kennedy took full responsibility for the Bay of Pigs fiasco, his approval ratings *increased*, leading him to quip, "The worse I do, the more popular I get."

Ask for help

As a Vice Admiral, Tom Collins was the Commander of the U.S. Coast Guard's Pacific Area, a region of responsibility that encompasses 74 million square miles of ocean from the U.S. West Coast to Asia and from the Arctic to Antarctica. On February 4, 1999, the 640-foot freighter *New Carissa* ran aground on North Spit Beach in North Bend, Oregon. The ship began leaking fuel oil and diesel onto the beach and into the water, eventually totaling approximately 70,000 gallons; or enough to cover about 65 acres.

> We had a partnership and a response plan that we had worked with the state and local [authorities], and we implemented that to deal with that crisis. The first attempt was to try to burn it. This is pristine, beautiful environment. We tried to burn it, and there's a lot of oil. That stuff, believe or not, it's hard to burn. There's separate compartments [and] void spaces [in a ship]. [...] It's difficult to get on fire and burn it off. That was unsuccessful, and then it broke in half, and we finally, in coordination with the state environmental folks, said, "Let's pull it off the beach, take it offshore into really deep [water]." There's some really, really deep water off there. "Let's sink it, and it'll go to the bottom. The temperatures at the bottom will keep it congealed. There's nothing that it can hurt out there."

The plan was then to take tugboats and tow the freighter out to the deep water, but then the question they faced was how to sink it. Ships are more difficult to sink than might be imagined because they have numerous void spaces throughout them where pockets of air collect, which can keep the ship from sinking completely. As he wrestled with this question, then-Vice Admiral Collins realized he needed help from someone with expertise in sinking ships. As the district commander in Hawai'i, Admiral Collins had built partnerships with personnel in the Department of Defense, especially within the Navy, and one of those was Vice Admiral Denny McGinn, Commander of the U.S. Navy Third Fleet in San Diego.

> I called a partner, [...] Vice Admiral Denny McGinn, [...] and I said, "Denny, I've got a problem. Can you help me? I need this thing to be sunk offshore. You probably have the capability to do that." He said, "Oh, you know, Tom, very well that sometimes these things don't go down easy. But if I take this job, it's going down. I can't see in the headline, 'Navy unable to sink ship.' That headline is not good." He mobilized a surface combatant and a sub, and they used both, put munitions into it, and sunk it.

A part of doing our best is recognizing we cannot do everything ourselves. If we are to do our best, we must ask others for help. Just as many people are afraid to admit and own their mistakes, many leaders avoid asking for help for fear of being seen as inadequate. Getting the

★

right people to do the job is excellent leadership, not ineptitude. Excellent leaders get the job done through people, not by themselves.

HOW SHOULD YOU RESPOND TO A PROBLEM?

When he was in Vietnam, then-Captain Johnnie Wilson saw firsthand how a lack of unit cohesion caused problems for the Army. That lack of unity resulted in soldiers being injured who otherwise would not have been, and Captain Wilson realized it was vital as a leader to "understand that I have this mixture, this diverse platoon, and I've got to be careful and treat them all the same and care for them all the same."

After Vietnam, when he was a battalion commander at Fort Lewis, then-Lieutenant Colonel Wilson was training with his battalion at Yakima Training Center when he recognized the same problem he had seen in Vietnam.

> We had to go out and bivouac. So, one night about 2200, 2300, I was walking through the area looking at some of the tents. So, here was this GP large (general purpose large tent), and I walked in, and every soldier in that tent was African American. I said, "Oh, hell no." Because in this platoon, there have to be welders and mechanical maintenance and POL (petroleum, oil, and lubricants) guys. So, I got the sergeant major, and we had a meeting right there, and I said, "We've got an hour to get everyone where they belong."
>
> If you don't stop those kinds of things, then all you're going to do is just [keep] perpetuating that. If you do that and then they're in their particular platoon and they have to complete a mission, [because of a lack of cohesion] you [won't] have them working together timely on refuel on the move [or] re-arm on the move. Everyone, just like [on a] football team, has a role to play. So, [...] for the sake of the organization, you can't afford to have people operating outside of where they should be, because we've all got to work. We've got a doggone mission to do. We've got a role to do, and if we're not there when needed, it's not going to happen.

General Wilson was not alone in his insistence that a leader should never walk by a problem; over half of the Four-Stars said the same thing. Whether we want to be or not, leaders are held up as *the* example by those we lead. When something goes right, people look to the

★ ☆ ☆ ☆

leader to gauge their response. When things go wrong, they look to them to determine how wrong. Leaders become the standard—what they accept becomes acceptable, and what they reject becomes unacceptable. As a leader, if you see a problem and walk by it, ignoring that problem becomes acceptable. General Ann Dunwoody said, "If you walk by a mistake, whether someone's not wearing their headgear [or something else], you just set a new lower standard." Similarly, General Bob Kehler said, "If I walk by a problem, everybody else will accept it." As the leader, you have the responsibility to uphold the standard and to never walk by a problem. If you repeatedly ignore the standard, perpetually lowering it, eventually there will be no standard, and things will be in disarray. In addition to lowering the standard, when a leader walks by a problem, it conveys one of three things shown in Table 2.1.

Table 2.1. The Impact When a Leader Walks by a Problem

WHAT DOES WALKING BY A PROBLEM CONVEY?	WHAT DO PEOPLE THINK AS A RESULT?
It isn't a problem.	It is okay to ignore the problem.
The leader didn't see the problem.	The leader is either not paying attention or does not know enough to know it's a problem.
The leader doesn't care about the problem.	The leader doesn't care about the standard, the group, or their own role as a leader.

No matter what the reason is for why you walk by a problem, doing so is clear evidence of poor leadership. From the Four-Stars' discussions, I identified two crucial practices for avoiding walking by problems: setting and maintaining a high standard and replacing people who are impeding success.

★

Set and maintain a high standard

As a lieutenant colonel, Gus Perna was a battalion commander. Shortly before a large-scale training exercise, a young officer was assigned to his staff. The officer had previously served on the staff of a very high-ranking officer. Consequently, the officer had become accustomed to less physically demanding activities and a less rigorous operating standard.

> My staff is about warfighting. We're a battalion. We're going to go [and] be in the middle of the fight. Standards and discipline were really important to me; doing the mission to its requirement was important. You have to be good at what you do, to be agile, adaptive, innovative. [...] On our first field training exercise [for] the battalion, it just wasn't where it needed to be, I'll just leave it like that. So, I was in the constant, "Okay, that was a great effort. Redo it. Great effort, redo it. Great effort, redo it." I just kept saying, "Not good enough, do it over again. Do it over again." My philosophy was, when you're out there everybody is in the priorities of work, because if the enemy comes over the hill, it's mildly interesting that the officers don't do work or somebody's in charge of this or that. They're going to go for the weakest spot. So, it was a constant do over, it was redo-redo-redo.

In the midst of the constant repetitions of the exercises, then-Lieutenant Colonel Perna's insistence on achieving the standard became too much for the new, young officer to bear.

> It was too much, about the third or fourth redo. [...] I sent a sergeant major over to [...] give [the officer] a pep talk and get [the officer] back in the fight. Then later on, I [took the officer aside], and I said, "Hey, let's talk about what happened here. Let's talk about our profession. Let's talk

★☆☆☆

about what we're getting ready for. Let's talk about what is the worst thing that could happen to me, [which] is that you're not ready and then you die. That's my responsibility. That's what I'm accountable for, number one. Number two, you're not ready and you fail at your job because you're tired, because you didn't do PT (physical training), because you didn't set up the standard, and now I'm out an intelligence officer."

After demonstrating the high standard to the young officer and explaining his reasons for maintaining it, General Perna surmised the officer "ended up being one of the best leaders that I ever had on my staffs; [they were] just tremendous."

Maintaining a high standard isn't just serving as a philosophical exemplar; it may well be a matter of life and death. General Tom Hill said, "You've got to set a standard, and you've got to maintain the standard and be vigilant about it. Anyone who's been in direct combat has at least one person who's dead because they didn't follow that rule all the time. I've got one. [...] every now and then his face will cross my brain, and it's like it was just yesterday. He's dead today because I know I did not do, at that moment, my job right in terms of enforcing standards." Military contexts aren't the only place where maintaining a high standard can be of vital importance; there are myriad fields where high standards are imperative. From operating rooms to vehicle manufacturing to water treatment to the airline industry to commercial crops and beyond, failing to maintain the standard has led to countless deaths. We leaders can avoid that by exemplifying the importance of high standards and instilling that in our people.

Relieve people who are impeding success

When Hurricane Katrina devastated the Gulf Coast in August 2005, then-Vice Admiral Thad Allen was the U.S. Coast Guard Chief of Staff and was placed in charge of the search-and-rescue and recovery efforts. In addition to overseeing the large-scale efforts along the Gulf Coast, he had to give frequent briefings and interviews to governmental officials, reporters, and news outlets. During an interview with a news channel regarding the mass evacuation of people from New Orleans to Houston and San Antonio, an attorney general asked Admiral Allen

★ ⋆ ⋆ ⋆ ⋆

why the Federal Emergency Management Agency (FEMA) was not releasing the names of registered sex offenders who might be evacuated to those cities.

> I had never even heard it was an issue. I had never even thought about it, and I was live on television. So, I basically said, "I was not aware it was an issue. I'll look into it immediately, and I'll either issue a statement, or you can interview me again tomorrow, but I'll get to the bottom, and I'll get back to you."

Immediately following the interview, Admiral Allen set out to investigate the attorney general's question. He asked the attorney assigned by FEMA to support him to find out if there was a way they could determine whether evacuees had any kind of criminal record about which there might be concerns.

> He came back about 10 minutes later, and he said, "Well, when you apply for FEMA benefits or we're assisting you, there's privacy act considerations, and I don't think we can give him that information." And I said, "That doesn't sound right. Go back and ask again." So, he came back a few minutes later, and he goes, "Listen, we can't give him that information." I looked at him and I said, "Wrong answer."

Recognizing he could not rely on the attorney from FEMA, Admiral Allen made a couple of calls to the Department of Homeland Security and ultimately spoke with the general counsel there who told him how to make the appropriate request to get the information. "Problem solved," he said. "So, I went back, and that's what we did. I made the announcement. It ended up not being an issue, but we answered with credibility, didn't try and stretch the truth, or anything."

A week later, two people from the Centers for Disease Control and Prevention (CDC) walked into Admiral Allen's office and asked him if they could try to determine who in New Orleans had tuberculosis and to where they might have gone in the midst of evacuations and mass housings. They wanted to try to prevent an outbreak.

> I said, "Of course." So, I called the same lawyer, and he went away for a while and he came back in, and he said, "Oh, you can't give him that information. It's privacy and health information and everything else." I said,

★ ☆ ☆ ☆

"You're fired." He said, "What?" I said, "You're fired." He [replied], "I work for FEMA; you can't fire me." I said, "Oh, that's right. I don't have any authority to do that. Leave. Get out of my office. Don't come back. I can do that."

Admiral Allen knew that there is mandatory reporting for people who have certain communicable diseases, and tuberculosis is among them. The attorney's refusal to work to provide that information to the personnel from the CDC meant he either did not know the law, did not want to do his job, was completely unaware of the public health ramifications of the exposure of large groups of displaced people (e.g., in the Super Dome) to others with active tuberculosis, or some combination of the three. Admiral Allen identified that, and the gravity of the situation, and swiftly relieved him of his duties.

When I asked about a leadership lesson they learned the hard way, the most common response from the Four-Stars was leaving someone in a position when it was clear they were not performing up to standard and were impeding organizational success. Over one-quarter of Four-Stars spoke specifically about the importance of relieving people who are not performing well. General Joe Ralston said, "When somebody that worked for me was not performing up to the standards that I thought that they should be, I would have a tendency to call them in and say, 'You need to improve here, here, and here. I'm going to give you 60 days to do it, or I'm going to give you 90 days to do it, but I want you to work hard on that.' Invariably, at the end of the 60 days or the 90 days, I would wind up firing them anyway and hiring a replacement. The lesson that I got from that is the replacements were always better. I finally decided, 'When I decided somebody was not hacking it, go ahead and remove them then, and get on with it.' Nobody likes to fire somebody, [but for the sake of the] organization, you need to do it, and you need to do it sooner rather than later."

It can be challenging for leaders to have to remove people who are failing to perform, whether from a lack of capability or character. As leaders, we often develop relationships with those people, and we may have been instrumental in those people being chosen to serve in those roles. When either is the case, we become biased—we want

★

to believe we made the right decision in hiring them. We can lose our ability to be objective. We look for things to confirm our original decision to hire them and ignore their shortcomings (i.e., confirmation bias). Additionally, if we have a relationship with someone, we want to believe the best about them. General Gus Perna recalled the difficulty he faced when he had to relieve someone.

> If you're not careful, that bond can help you or can handicap you to making the right decision. As a Four-Star, I had to relieve a commander who I had really a lot of respect for. I don't want to fall into a trap again, but I had a lot of respect for that person. People are humans, and people make mistakes. It was the hardest thing I had to do; I lost a lot of sleep over it. I can clearly remember, I got on a plane and flew to the person's location. I walked in, and I relieved the person, and I walked out, and I had to get back on a two-hour plane ride because I was going to do it in person, and I had to sit there and freaking think about that, what I did. But [it was] what was best for the organization.

Failing to relieve people who are not performing up to the standard is, perhaps, the most egregious way a leader can walk by a problem, because it has far-reaching impacts on organizational morale, performance, and, in some cases, survival. Four-Star leaders make it a point to never walk by a problem, especially when the problem is that someone needs to be replaced. Excellent leaders serve as the example—through doing what is right, doing their best, and never walking by a problem—because their character demands nothing less.

★ ☆ ☆ ☆

CHAPTER 3

SERVICE OVER SELF

[Leadership] is a selfless calling, and the sooner you realize that, particularly the selfless part, the better off you're going to be and the more successful you're going to be.

– GENERAL JIM JONES, U.S. MARINE CORPS –

In October 2007, then-Brigadier General Vince Brooks was the Deputy Commander of the 1st Cavalry Division, the second in command in Baghdad during "The Surge." One evening he was in the base dining facility when a rocket landed at the entryway, detonating and killing two and wounding thirty others. The air was filled with smoke, dust, and the cries of the injured, and as the senior officer, Brigadier General Brooks worked to clear his mind to lead his troops in the chaos. "First thought is, 'Okay, this is the "Saving Private Ryan moment" on the beach where the good captain is dazed, but his troops are looking to him and asking, 'What do we do now, sir?'" he remembered. "I guess it was my first thought of 'What should I do now?'"

The scene was chaotic, as soldiers outside of the facility repeatedly yelled for a medic, and the wounded were being carried into the dining facility. General Brooks realized that the best thing he could do was physically to serve, to the best of his abilities, those wounded soldiers.

> It was going to take a few minutes before we could get the ambulances over there, and that would take multiple turns. Because there were 30-plus people, which we didn't know at the time, but multiple people

★

are injured. So, I grabbed one of them right away who had a penetrating wound into his lung, put him in the back of my vehicle, and carried him to the aid station and then started grabbing IV bags to try to help.

As the wounded began pouring into the aid station, though he could have assumed a position of command, General Brooks sought to serve the injured and those providing care to them.

Am I in charge when I walk in there, into that aid station, where people are significantly more junior to me and are very much overwhelmed by the inflow of patient, after patient, after patient, all of whom are in crisis? Some of them have traumatic amputations. Some of them are bleeding out. Some of them have mild lacerations, but they need care, nevertheless. All these things are happening all at the same time. So, I found myself not only evacuating that one soldier, but being inside of the emergency area there at the aid station, trying to help one particular soldier who was in a very, very bad way. He didn't look like it. He had just two small penetrating wounds, and he was dying; there was no doubt. He had injuries to his inside that we couldn't see. We're trying to find a way to save him at that point in time. Literally, the best I could do was to hold an IV bag and support the specialists and the sergeants and the captains and the majors who were doing critical lifesaving measures, because they needed one more hand.

While the medical personnel rushed about to treat and save as many of the wounded as possible, Brigadier General Brooks stood at the side of the injured soldier's gurney holding the IV bag—a One-Star General serving as a human IV pole because that was what was needed by the medical team and the dying soldier. As the night wore on, General Brooks continued to serve in whatever way that was needed, regardless of his rank.

When it came time to evacuate, we still had multiple soldiers who needed to be treated. We had our first dust off helicopters coming in. This is the nighttime evacuation, and I need to carry the litter because I'm not an expert in what was going on inside of that emergency room, inside of that treatment facility, but I could help to carry somebody so that they didn't have to stop. They could get onto the next person.

★☆☆☆

As he reflected on his selfless service that night in Baghdad, General Brooks saw it as simply being a leader.

> All those things I think are examples of servant leadership, where I wasn't worried about my status. [As the Commanding General, I could've said], "As of right now, I'm in charge here. Do that. You do that. I don't know what you're doing, but do some more of it." That would've been a ridiculous thing for me to try to do, and would not have been leadership, but it was the best I could do just to be a servant.

Selfless service is acting on the willingness to do what needs to be done to accomplish the mission despite potential or actual personal costs. When discussed in the context of leadership, it is often referred to as "servant leadership," which was exemplified by Jesus Christ. However, the idea of servant leadership wasn't described formally until the 1970s when Robert Greenleaf did so, most notably through his booklet, *The Servant as Leader*. Many of the Four-Stars advocate for servant leadership, and about 40% expressly said that putting service of the mission and others ahead of yourself is a fundamental attribute of excellent leadership. While that commitment to selflessness may come as a surprise to some, it has been officially venerated in the organizational values of both the U.S. Army ("Selfless service") and U.S. Air Force ("Service over self"). Additionally, its value is storied throughout U.S. military history. A renowned example comes from General Dwight D. Eisenhower, the Supreme Allied Commander who led the D-Day invasion. During its preparations, Chief of Staff of the Army General George C. Marshall asked General Eisenhower what the principal quality was he looked for in choosing commanders. Without thinking, General Eisenhower replied, "Selflessness."

Service over self is a dedication to the greater good, embodying both humility and commitment to one's role. For military members, that means a dedication to the defense of our nation, specifically to the Constitution and the ideals it contains. This plays out in service to the nation as a whole and a commitment to the other members with whom one serves. However, the military is far from the only profession where service over self is a core value. Across diverse fields— such as firefighting, teaching, medicine, and even certain areas of

★ ☆ ☆ ☆

business—selfless service remains an integral part of the vocation. But the question arises, how does a leader develop the capacity for selfless service? Throughout the Four-Star interviews, I discovered that service over self manifests most often as one or more of four major precepts, which frame the remainder of this chapter.

WHAT DOES IT MEAN TO SERVE RATHER THAN BE SERVED?

When he was a colonel, Keith Alexander was a brigade commander in the XVIII Airborne Corps at Fort Bragg. Individual brigades within the Corps would take turns running a jump session one Saturday every month, during which the families of the soldiers would sell concessions to raise money for the wives' clubs, to be used to support troops in need. Each brigade's families would prepare for weeks in advance of their upcoming jump session. Once, immediately before the Engineer Brigade's weekend, a hurricane hit Fort Bragg, and their ability to run the jump session was suddenly in question.

> A hurricane came through, knocked down several hundred trees across Bragg, and the engineering unit had to go fix those. [That brigade] was responsible for running the jump that weekend. [Speaking to the Engineer Brigade commander], the deputy corps commander said, "Jack, you can't do it. You've got to take care of the trees." [The Engineer Brigade commander replied], "But sir, the families have already gotten this all done. We're all set." [But] he said, "I'm sorry, we're going to have to give this to another brigade. [...] Who would like to take it? Now you've got four days to prepare."

This presented a difficult predicament for the other brigade commanders because running the jump sessions took a lot of extra work. Additionally, the families wouldn't have time to do all of the necessary preparations to have concession sales, which meant they would lose money. So, as General Alexander recalled, no one wanted to take on the duty.

> It was all of a sudden, you look around, everybody's head is down. You think, "Well, that's crazy." I said, "I'll take it, and more importantly, Jack's families can run the stuff. It's okay. We'll have the families here, and we'll run [the jump session] for them."

★ ☆ ☆ ☆

Then-Colonel Alexander's brigade took on all the extra work of running the jump session, while still allowing the Engineer Brigade's families to make the money from concession sales. Recalling that story, General Alexander said, "And they never forgot that."

In many Western societies, people view the leader as being elevated above those being led. This can be seen across various settings, from the way kings and queens are extolled to the physical position of the U.S. President when delivering the State of the Union address. There is a general idea that being a leader means being above everyone else and receiving the benefits of that exalted position. This outlook has even been codified in the vernacular of the military: R.H.I.P.—"rank has its privileges." Unfortunately, with this ingrained societal perception of leaders' position, we may begin to view ourselves in the same way— above those we lead. When that happens, it is only a matter of time until we slip into abusing the power of our positions to serve us instead of serving those we are entrusted to lead. This human tendency is what led Lord Acton to say, "Power tends to corrupt. Absolute power corrupts absolutely." No matter how certain we are of our own rectitude, we must vigilantly guard against elevating ourselves above those we lead. Capitalizing on one's leadership position for personal gain is at best terrible leadership and at worst, perhaps, megalomaniacal and sociopathic.

Leadership has nothing to do with your position and everything to do with who you are. In a prior research study, I analyzed the responses of other experts in leadership and found that the heart of leadership is "influencing others by means of coaching their capabilities, ensuring their resilience and performance, and serving as an effective role model." There is nothing in that definition reliant on one's position of power or authority. It is a way of being, one dedicated to the accomplishment of the mission and service to others—those we lead. That way of being is based in our character. If we can keep these things in mind, it will help inoculate us against being corrupted by our position and will empower us to serve the mission and those we lead ahead of ourselves. In interview after interview, the Four-Stars emphasized that service over self is essential to leadership and that it requires a

★ ★ ★ ★

conscious commitment because it requires humility, can be difficult, and is often not glamorous.

As humans, we have a lot of things that make us want to serve ourselves, such as ego, self-gratification, and self-preservation. From an individual biological perspective, those drivers work to ensure that we live long enough to propagate our DNA to the next generation. That biological drive gives us a tendency to put ourselves at the center of our prioritization system. Additionally, as a species, we are historically relatively weak and have a lot of natural predators, factors that anthropologists attribute to our highly social nature and our drive to value and help others. However, when it comes to mission accomplishment, we do not necessarily have the same sorts of innate drive. So, our biological programming is to prioritize ourselves over others and others over the mission (Figure 3.1). However, Four-Star leadership prioritizes things in the opposite order, placing service to the mission and others over self.

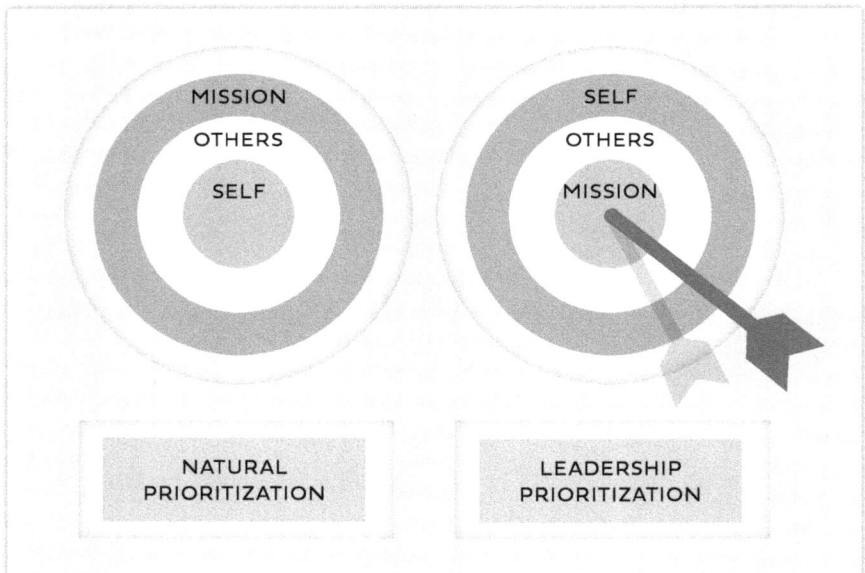

Figure 3.1. Four-Star Leadership Prioritization: Service to Mission is the Center of the Target. The natural human prioritization is to place self at the center of the bullseye, with others next, and the mission last. Appropriate leadership prioritization is to place the accomplishment of the mission at the center of the bullseye, with others next, and finally self.

★ ☆ ☆ ☆

Since prioritizing the mission above others and self contradicts our instinct, doing so can raise issues with those who are not thinking about the reason for the group's existence—the mission. Multiple Four-Stars emphasized the central position that mission accomplishment must take, and they followed it closely with service of those we lead. Admiral Mike Rogers said, "You will find some who always [say], 'Well, it's all about people.'" Without minimizing the importance of people, Admiral Rogers always answered them, "We only exist for one reason, and that's to execute the mission. Guys, priority number one is the mission." He went on to say, "Now, a good leader knows you can't do that without well-led, motivated individuals who understand the intent, who understand their role, who understand the vision." Similarly, General Bob Kehler recalled having to address this point with his team, telling them, "The mission is the most important thing." He said, "I would hear people say, 'I put my people first.'" But he explained to them that wasn't the case, saying, "No, you don't. If you do, you will never get the mission done because putting your people first means you never send them outside the wire. You can't put your people first. You've got to put the mission first, but you have to advocate for your people all the time, and you have to take care of their needs." Fundamentally, without a mission, without a goal, without something to work toward as a group, there is no need for leadership. People do not need a leader to stay where they are and accomplish nothing. The goal—the mission—establishes the need for leadership, and everything else follows from that.

SELF REFLECTION

If asked anonymously without threat of coercion, would those you lead describe you as someone who seeks to serve rather than be served? What proof is there for or against that?

★ ☆ ☆ ☆

WHAT DOES IT MEAN TO HAVE MORAL COURAGE?

As an Army Ranger who served as a Special Forces operative, General Stanley McChrystal demonstrated more than his fair share of physical courage throughout his career. His exemplary service saw him rise to become the Commander of the Joint Special Operations Command—leading the U.S. military's most elite and secretive Special Forces operators. When he later was the Commander of U.S. Forces in Afghanistan, General McChrystal was faced with a challenging situation that required great courage. As it turned out, while there he learned a leadership lesson the hard way.

> There was an officer who was the best battalion commander in Afghanistan. I was a Four-Star, he was a lieutenant colonel, and I knew him; he'd worked for me before. He was the best battalion commander on the ground in the country. I cared a lot about him as a human being, as well as a commander. Well, he had this morning briefing, and in the morning briefing he used these humorous meme photographs.

Though the memes in the briefings weren't pornographic, General McChrystal surmised that while they were humorous, some of them could be judged as slightly sexist. As a result of the memes, someone filed a complaint with the battalion commander's boss, the brigade commander.

> His brigade commander didn't like him anyway, so his brigade commander recommended [that the battalion commander] be relieved [of duty]. His division commander accepted that recommendation and came to me and said, "We've got to relieve this guy." I was caught on the horns of the dilemma. The division commander was a good friend of mine, and I trusted him, and I wanted to support him, but I thought his judgment in this case was wrong. Yet clearly, if I overrode it because the guy (the lieutenant colonel) used to work for me, it was going to look exactly like patronage. So, I didn't. I let the officer be relieved of command, it ended his military career. [...] I let an officer's career get destroyed that shouldn't have been, and I did it because I didn't have enough courage to separate what the greater right was, what the greater justice was. [...] to this day, I don't think that I let my values actually run that. I let my conventionality drive that. So, I regret it.

★ ☆ ☆ ☆

Putting service over self can be excruciatingly difficult. For some people in certain fields, it may require risking life and limb. Not all of us are faced with such challenges. Yet, we all face moments where a certain type of courage is demanded of us if we are to put serving the mission and others ahead of ourselves.

There are several definitions of the word courage, most of which could be distilled into "the determination to accomplish a goal in the face of danger and/or fear." Usually when we think of courage, we think of the physical courage of those, like Special Forces operators, who perform acts of daring. Whether it is jumping out of an airplane, fighting off a predator, or rushing into a burning building, the mental picture that is evoked by the word courage is usually rooted in something physical. However, General McChrystal's story shows us there's more to courage than intrepid physical feats.

What General McChrystal described wasn't life-threatening. It was more like the type of moment that everyone faces at varying points in our lives. The courage required wasn't physical in nature, but rather it was what is typically referred to as moral courage. Moral courage is different from physical courage. Unlike the image presented in the previous paragraph, moral courage could be defined as the determination to do the right thing in the presence of personal psychological, social, emotional, or financial risk. Because it is predicated on the ability to identify and do the right thing, moral courage is deeply rooted in one's character. This relationship between character and moral courage was acknowledged by several Four-Stars. For example, General George Casey said, "character and courage are [...] closely entwined." Similarly, General Ann Dunwoody equated character to moral courage when she said leaders of character "have the personal courage to do the right thing for the right reason." Beyond the importance of character to moral courage, most models include fear as an element of it, and that fear is specifically of the consequences of the required action. So, we can visualize moral courage as the confluence among determination, a goal (to do the right thing), and fear of the consequences of doing the right thing.

★

Figure 3.2. Moral Courage. Moral courage occurs when someone has the determination to pursue a morally-based goal (the right thing) in the face of fear of the consequences of both the pursuit and accomplishment of the goal.

The role of moral courage in leadership was a common theme for most of the Four-Stars. Though they had faced innumerable times that called for physical courage, they view moral courage as far more important for leaders. This is because time after time, their leadership was instead tested by psychological, social, emotional, and financial issues, as well as the consequences of their decisions. Those decisions that required moral courage were not about their own personal physical safety; they almost always were accompanied by risk and uncertainty for themselves, others, and our nation. General George Casey said for leaders to succeed, "It takes courage to act with conviction in the face of uncertainty and risk. Nothing good happens without risk, and that's where the courage comes in, and it's only getting more uncertain today," given the increasing social, cultural, and political volatility. Four-Star leadership requires moral courage.

Being a person of character, having moral courage, and doing the right thing is often hard. The reason for that should be obvious—it does not take exceptional character or courage to do what is easy, what is expedient, and what is free of risk. Instead, it takes courage to do

★★★★

what is hard, what is unpopular yet right, and what carries personal risk. This explains how having the moral courage to support the lieutenant colonel was so difficult for General McChrystal; in fact, this act of moral courage was even harder than the physical courage required of an Army Ranger expected to parachute from airplanes, rappel from helicopters, and storm enemy fortifications. Situations like the one General McChrystal faced—that is, supporting someone he was leading—are perhaps the most common type requiring a leader to have moral courage. What a leader does in such situations becomes highly important to their subsequent ability to lead.

Those we lead are always watching us, which means we must always have the moral courage to do the right thing. General Tony Zinni said, "Even when you don't think people are watching you or paying attention to you, they are." According to General James Conway, they are watching "how you make decisions [and] how you handle issues." They are always watching us to determine if we will do the right thing, and this is especially true when it comes to backing our people—in other words, them. Few things can faster destroy your ability to lead people than not supporting them when they've done what you asked and something bad happened as a result. Unfortunately, it is not uncommon for people in leadership positions to pass the blame and try to protect themselves (see untold numbers of political figures as examples). Passing the blame is an utter failure of moral courage, a complete absence of it. If we do that as leaders, our people will see our lack of moral courage—how we abandoned them when things got rough—and they will no longer trust us, will give up on us, and our ability to lead will collapse.

Be willing to be fired

As the Director of the Ballistic Missile Defense Organization, then-Lieutenant General Les Lyles was summoned to a hearing before Congress to testify regarding the U.S. missile defense system, and he knew he would be facing challenging questions.

★ ☆ ☆ ☆

During one congressional hearing, we had a particular strong zealot [...] a congressman who was a very strong supporter for missile defense. [...] He was a very, very strong congressman. He had a very strong position where he had oversight of funds for the military. [...] For some reason, he had the impression from his staff that I was going to answer a certain question in a way that would allow him to embarrass President Clinton by saying, "See, Clinton is not putting any money into this particular program, and he should because of its great success." But I did not give that impression to him or his staff.

When it came to that hearing, I vowed to always tell the truth, whatever the answer might be. So, when he asked me about adding money to this particular program in terms of literally scores of millions of dollars, [...] I basically said, "Congressman, if you give me that money for this program, it'll be like throwing good money after bad, because there are still technology problems, and I have to solve them." He took that as my saying that he was wasting money pouring it into missile defense, and he stopped the hearing right there, literally stopped the hearing, came down from his perch as the Chair of that particular hearing, came up to me sitting at the witness table and said, "I can't trust you," because he thought I was going to say what he wanted me to say. My immediate rejoinder to him was, "Congressman, if you can't trust us in uniform, who can you trust?" He walked away, [and] the hearing ended. He was a powerful congressman. What I thought right away was, "Well, that's the end of my career because I told the truth. But, this congressman didn't appreciate the truth, and he's so powerful, it's going to end up ending my career."

The day following the episode with the congressman, Secretary of Defense Bill Cohen called General Lyles and told him, "Don't you worry about what happened yesterday. You did exactly the right thing. You told the truth. Don't worry about that congressman. [...] Me and my friends on The Hill will take care of him." Subsequently, the congressman apologized to General Lyles for his behavior, and the General received a lot of support from others for his courage to tell the congressman the truth despite the risk to his career. "In my meetings with the corporate leadership [and] all the Four-Stars, I was immediately greeted by them all, thanking me for the way I handled the situation," General Lyles recalled. "I actually attribute that to one of the reasons why I got promoted to my fourth star."

★ ☆ ☆ ☆

When discussing selfless service, some of the Four-Stars referred to the need for leaders to have a willingness to be fired. In saying that, they were not implying that a leader should be cavalier or arrogant. Instead, they were describing the determination to do the right thing when it is accompanied by a significant personal professional threat, a combination that expressly demonstrates moral courage and character. It is a commitment to discharging your duty in spite of the possible consequences. General George Casey tied the willingness to be fired back to the commitment to service over self.

> It's hard to get anybody to disagree with their boss. It takes a level of courage, at any level, to disagree with your boss. When your boss is the President of the United States, and you're a serving military officer who's been respecting the Commander-in-Chief for 40 years, it's really hard. So, it just takes that level of commitment to something larger than yourself.

Generals Casey and Lyles weren't the only Four-Stars who used the willingness to be fired as an example of the character required for excellent leadership. Whether telling the unwanted truth to the President, members of Congress, higher ranking officers, or foreign dignitaries, the willingness to be fired was a common theme, and Admiral Thad Allen suggested that the demand for that willingness only increased with increasing responsibility. As he thought about when he led the responses for Hurricane Katrina and the Deepwater Horizon disaster, he said, "You should never take a job like that unless you're prepared to be fired."

General Mike Scaparrotti, *on Being Willing to Be Fired*

General Mike Scaparrotti describes one such circumstance from his own career.

> We were in a brigade that had a quite talented brigade commander. He was recognized in the Army as probably one of the up and coming. But he, too, was so focused on his own advancement. [He] was being abusive to company commanders. What I mean by that was, he'd put his hands on people and grab 'em, one-on-one and threatened them, et cetera. I didn't know this until I worked directly for him.

★

The brigade commander's behavior continued and was wearing on soldiers throughout the brigade, especially the junior officers. Finally, one battalion commander who worked for the brigade commander confronted him on the problem.

> [He said], "That's it, sir. Enough of this crap." He actually risked his career telling this guy that. He did the right thing. Of the battalion commanders, he was the only one that was willing to come forward and say, "Look, we all know what's going on, and this is just not acceptable." That battalion commander set an example for me. You've got to be willing to speak up. You've got to do what's right.

The willingness to be fired demonstrates dedication to doing the right thing, which may mean doing or saying things that people with power do not want. In that circumstance, service over self means not being so enamored with your job or position that you compromise on what is right in hopes of protecting your role. If you compromise to protect your position, you may appease someone in power now, but people— both those you are trying to appease and those you have been charged to lead—will see you as weak in character and untrustworthy. That will undermine your ability to lead others, and it will catch up with you, likely sooner rather than later.

Across their careers, there were multiple ways the Four-Stars demonstrated a willingness to be fired. The most described circumstance was speaking truth to power—telling someone who is in a higher position of authority what they do not want to hear, when they have the power to have significant negative impacts on one's professional, social, and financial life. The Four-Stars' willingness to be fired demonstrated strong character and a firm commitment to service over self, whether they were taking a necessary action that might get them fired, refusing to follow an immoral or illegal order, standing up for a subordinate who failed while carrying out their instructions, or speaking truth to power.

Speak truth to power

As General Pete Chiarelli discussed what is required for successful leadership, he said, "You've got to be willing to speak truth to power,

★☆☆☆

and that can be extremely, extremely difficult, no matter what level you are." He then shared an excerpt from a 2022 *New York Times* book review by Michael Shear.

> The experts made clear that Mr. Trump did not always get the yes man that he wanted. During one Oval Office exchange, Mr. Trump asked General Paul Selva, an Air Force officer and the Vice Chairman of the Joint Chiefs of Staff, what he thought about the President's desire for a military parade through the nation's capital on the Fourth of July. General Selva's response [...] was blunt and not what the President wanted to hear...

General Chiarelli then quoted from the article to describe what General Selva had told Mr. Trump.

> I didn't grow up in the United States, I actually grew up in Portugal. Portugal was a dictatorship, and parades were about showing the people who had the guns, and in this country, we don't do that. It's not who we are.

General Chiarelli finished reading the excerpt and expounded on the importance of leaders following General Selva's example.

> Now that's speaking truth to power. That's hard to do, particularly to a guy like Trump who really doesn't want to hear anything other than words that reaffirm the decision that he's already made. So that's a really, really difficult thing for folks to do, and you need to get started on it earlier in your career, whether it's the first lieutenant to the XO (executive officer) or to the company commander who's about ready to make a really stupid decision. You've got to be willing to stand up and speak truth to power. Then you've got to realize that if the individual doesn't listen and the order is not illegal or immoral, you are duty-bound to go ahead and do what you've been ordered to do. But, I think it's absolutely critical that you speak truth to power.

Speaking truth to power is arguably the most difficult demonstration of moral courage and, therefore, one of the highest forms of selfless service. That is because speaking truth to power is a direct, undeniable refutation of the person in power. Whether you take an action you were told not to or you avoid an action you were ordered to take, that decision and action are almost always remote from the person in power—your action is an indirect response to the leader's authority. Additionally, if you are defending a subordinate who failed as a result

★

of doing something you told them to do, most people will at least credit you with defending someone in a weaker position, which may soften the repercussions. However, when you directly speak the truth to someone in power, it is often taken as an affront to the person's position and authority, a direct challenge to the person's face. Doing so requires taking a significant personal risk—we could lose respect, lose our job, or in the most extreme circumstances, lose our life. It takes an extraordinary leader to respond positively to being challenged like that. Unfortunately, many people in leadership positions do not have the wherewithal to respond positively and, instead, return what they see as an attack with full force. Most people recognize this risk, which is why it is so hard to do and rare to see people speaking truth to power.

In addition to being willing and able to speak truth to power, it is vital for leaders to have people who are willing to speak truth to them. General Charles Krulak spoke about this, saying, "The more senior you get, the more important it is that you surround yourself with people with moral courage [who are] willing to tell the emperor or the emperors that they aren't wearing any clothes. I've found that to be critical. It's critical when you display it, [and] it's critical when you surround yourself with people who would display it to you. It keeps you out of trouble."

General Jim Mattis, *on The Impact When We Don't Speak Truth to Power*

A month removed from September 11, 2001, Operation Enduring Freedom began in Afghanistan, and then-Brigadier General Jim Mattis was the Commander of the 1st Marine Expeditionary Brigade. Prior to taking the role, he had been under the command of a Navy Admiral. He said, "I knew my Navy commander; we'd been ashore [together]. Navy Commanders give you a lot of freedom" to make decisions and take action. Once in Afghanistan, Mattis had to report to the Army General in charge of U.S. Central Command (CENTCOM), whom he did not know well. Nevertheless, he knew they were on the same page regarding the mission: find and eliminate Osama bin Laden.

★ ☆ ☆ ☆

General Mattis recounted, "Shortly after we'd arrived there, probably within the first 60 days in Afghanistan after 9/11, our intel people knew where Osama bin Laden was. They'd found one of two valleys, and he was making slow progress; he couldn't make fast progress with our command of the air toward the Pakistani border. As soon as I heard it, we swung into action." General Mattis intended "to take the assault troops up and" eliminate Osama bin Laden. He said, "I knew the ferocity and the skill of my sailors and Marines. When we closed in on the enemy, my only concern was how many boys was I going to lose? I knew we would win."

Because of a need for supplies, there was a brief delay in launching the mission. He explained, "Because it was going to be a lot colder where we were going, and the air a lot more rarefied, we needed a different kind of cold weather gear," which had to be flown in from Navy ships off the Pakistani coast. The sailors aboard those ships "already had the cold weather gear brought out of the various parts of the ship or down in the holds, had it ready, and they were loading it up onto helicopters to take it to the beach where KC130s were being flown in from Jalalabad." Meanwhile, in preparation for the impending operation, General Mattis' Marines "were pulled off patrols, cleaning weapons, getting food, getting warmed up, because they were going in."

In the flurry of activity and command, as they were working through the night to launch the mission, "All of a sudden, we got word to stand down (cease the operation to attack bin Laden)," General Mattis said. When he asked the reason, the response was "CENTCOM wants to use local [Afghan] troops," perhaps in an effort to improve the political optics of the situation. Though he tried to persuade the CENTCOM commander that time was of the essence and help him understand that most of the Afghans hated the group against whom the attack would be made, the response remained the same: stand down. "Bottom line: he got away," General Mattis recounted, "and it took 10 years before the SEALs caught up with him."

On reflecting on the profound historical significance of the order that he and his Marines should "stand down," General Mattis said his error was that he took for granted that he and the Army General were

★ ☆ ☆ ☆

on the same page regarding the mission and how it should be conducted. "I was not keeping my higher ups as fully informed as I should have," he said. "That is a mistake. That is a lack of mature understanding of a leader's responsibility." He had taken for granted that the Army General would operate like the Navy Admiral, giving him the authority to conduct operations as he saw fit. In retrospect, he saw "The Army likes everything staffed (discussed with superiors)," because they fight in different environments and worry about different things. As he thought about those events, General Mattis determined he should've been more insistent with the CENTCOM commander, should've spoken truth to power more forcefully. And if that didn't work, he should've gone further up the chain of command, because the situation was too important not to. Solemnly reflecting on those events, General Mattis said, "I failed," and "it cost our country."

While rare in his ability to own it, General Mattis is not alone in failing to speak truth to power. In professional, political, social, and family relationships, most of us fall prey to it often. We may halfheartedly mention something but give up on pursuing it if there is pushback. It can be difficult to muster the moral courage to overcome our fear of the consequences. Four-Star leaders put service to the mission and others over themselves. Maintaining that prioritization helps in summoning the moral courage needed to speak truth to power. In the more than two decades since those events in Afghanistan, General Mattis demonstrated repeatedly a dedication to service over self and an ability to speak truth to power. Undoubtedly, that was on no greater display than when he did so repeatedly during his service as the Secretary of Defense.

★ ☆ ☆ ☆

COMPETENCE

THE SECOND STAR OF LEADERSHIP

CHAPTER 4

COMPETENCE:
THE AGENCY OF LEADERSHIP

Leaders have got to have competence. That doesn't mean that if you're a house builder you know how to make the tools, but you know how to use the tools to build the house. That's the deal for a commander and a senior leader, in particular. You don't know how to fly the airplane, sail the submarine, drive the tank, shoot the mortar, do all that stuff, but you know how to put all those pieces together to get the job done, and you know how to motivate people to go get that job done.

– GENERAL BOB KEHLER, U.S. AIR FORCE –

On June 5, 2012, General Janet Wolfenbarger became the first female Four-Star General in the United States Air Force. Upon her promotion, she assumed command of Air Force Materiel Command, an organization of over 80,000 personnel, 75% of whom were civilians. As she stepped into the Four-Star position, the military was operating under the Budget Control Act of 2011, which had abruptly cut Department of Defense operations and maintenance (O&M). So, upon starting her new role as the first female Air Force Four-Star, General Wolfenbarger was thrust into determining how to continue all of Air Materiel Command's mission—each vitally important to the successful operations of the Air Force—with a budget that was 25% smaller.

★★

There was a directive that came down from the Secretary of Defense to all the Services, with a mandate to figure out efficiencies, kind of, "Anybody with good ideas, bring them forward," because we didn't want to stop doing the missions. There were very few missions that you would say weren't important that we had to keep doing. We just had to figure out how to do them with fewer resources, people, and dollars. So, that was going on when I took over as Materiel Command Commander.

To align with the requirements of the Budget Control Act and the orders of the Secretary of Defense, General Wolfenbarger had to take extreme actions to be able to continue the successful operations of Air Materiel Command.

When we had to take that drastic cut, what we ended up having to do— because you had to live within your means in the middle of a fiscal year— we actually cut the pay. We cut people's pay—the civilians'. [...] So, their pay was cut, and they were never reimbursed for the pay that they lost. It just astounds me. Sometimes when we can't get a budget authorized at the end of the year [and we are in] continuing resolution [where] people don't get paid, they always get back pay for what they should've been paid. That didn't happen this time. So huge, huge, tense, "burning platform," time to figure out how we could operate with far fewer resources.

With the major "burning platform" limiting her ability to continue the operations and maintenance of the Air Materiel Command, General Wolfenbarger had to identify other ways by which they could absorb the significant financial impact.

In the Air Force, Air Materiel Command is responsible for the science and technology mission, lifecycle management mission, developmental test and evaluation mission, and the sustainment mission, which included all depots and the 24/7/365 supply chain mission for the Air Force globally. Because of her prior assignments and expertise in how large segments of Air Materiel Command functioned, in her first few months of command she identified and implemented a tremendously positive streamlining of the Command's structure, beginning first with analyzing current practices.

We had 12 [subordinate] centers that previously had been doing [the four primary] missions, and it was really organized [in] a geographic fashion.

★★

So, whatever missions happened to reside at that geographically-located center, that center's commander oversaw. [That arrangement] precluded opportunities [for] sharing of best practices and lessons learned across each of these centers, because commanders did what you want commanders to do—they drew a circle around what they were responsible for; that was their priority. Their priority was not sharing. Their priority was executing their mission to the best of their ability.

General Wolfenbarger recognized the inefficiencies created by having the Command broken up across 12 subordinate centers. She determined that was a potential leverage point for reducing expenditures.

I took the opportunity to drive a huge reorganization of the Command. We went from 12 subordinate unit centers […] that were directly reported to the Four-Star, down to five […] [with a] huge reduction in overhead that resulted from that. So, $100 million bucks a year just because I didn't have to fund positions. No pink slips, I was quite proud of this; no pink slips, but still, [we] freed up all the budget that had been allocated for [those] 1,000 slots, [which equaled] $100 million on an annual basis. But that really wasn't the most powerful piece of this. What I did was align the way we were organized by mission set.

Having alleviated some of the demand to reduce expenditures, General Wolfenbarger turned more fully to streamlining the Command structure.

So, what we ended up doing was really breaking apart all that geographic alignment [and] put one commander in charge of each of the mission sets no matter where it operated, so I called it a geo-agnostic way to organize. For the first time in this Command's history, we [started] standardizing processes across all those different locations. Then, once you standardize the processes, now you can continuously improve them. You have to get after some standardization first to be able to realize where things are working well and where things aren't working well. So, we got after the best practices and the lessons learned in a more universal fashion.

As a result of the streamlining and process improvements that General Wolfenbarger introduced, Air Materiel Command saw tremendous reductions in expenditures and improvement in operations.

★★

After the first year, [...] you could start to see across each of those four primary mission sets [...] some great improvements. But we could validate that we had saved $3 billion, [...] either cost savings or cost avoidance, which means it wasn't yet budgeted, but it was going to have to be budgeted in order to execute our mission. [...] At the end of my three-year tenure, the Pentagon came in and did an independent assessment of the savings. They validated $6 billion after three years of either savings or avoiding. [...] That's exciting, the dollars that were saved, but can I say the most exciting thing for me was the mission effectiveness that resulted. Just one example: We had three depots. Each one of them, initially when I took over, had their own separate commander. We put one commander in charge of that sustainment mission [and started] figuring out how to produce aircraft out of a depot in a standardized way. We were delivering aircraft on these scheduled dates to the customers, to our warfighters (whereas, previously there were always delays). To me, that was even more exciting than the dollars that we saved. We were able to execute our missions just much more effectively.

Competence is having the knowledge, skills, and abilities to accomplish something successfully and efficiently. It is having what it takes to get the job done well and not necessarily being the person with the greatest expertise in a particular subject, which requires the personal character to be able to admit when we aren't the person with that expertise. For leaders, competence is instrumental in our ability to make informed decisions, inspire trust, and guide our teams to success. Leaders' competence has been associated with multiple indispensable components of successful leadership: trust and credibility, problem-solving, adaptability, innovation, decision-making, productivity, and long-term success. Knowing this, it is no wonder that nearly 60% of Four-Stars discussed and/or shared examples of competence as a key for leadership success.

It could be argued there is no greater crucible of leadership competence than combat. Afterall, the leadership task in combat is to accomplish a specific goal or attain an objective while at the same time attempting to prevent those being led from dying at the hands of an enemy actively seeking to kill them. Incompetence is swiftly punished with the loss of lives and, depending on the scale of the conflict, defeat

★★☆☆

and demise of societies and nations. That's the reason why so many studies of military personnel have found competence is *the* thing troops most frequently say they want in a leader.

After his second tour of duty in Vietnam, then-Captain Tony Zinni was put in charge of operating an advanced infantry tactics school for the Marine Corps. His students consisted primarily of Marines returning from Vietnam who had been quickly promoted to the rank of corporal or sergeant and, while having led squads, had never been through a training academy. Captain Zinni interviewed each of the students when they started at the school, and one of the questions he asked each of them was, "Think about your lieutenant, your platoon commander. You can only pick one quality. What is the one leadership quality that you would rank the highest?" In his interviews of hundreds of Marines, he found that 96% gave the same answer, "Competence… know his [stuff]." When General Zinni was later running corporations in the business world and conducting leadership research for a PhD, he found a similar result. "When an organization is under stress or in crisis, [a leader's competence] pops out as number one, because [the people in the organization say,] 'The boss has got to get me through it. He's got to know his stuff, or she's got to know her stuff.'" Conversely, he found, "When things are going well, you get a whole bunch of different answers. 'Oh, I want them to be caring.' 'I want them to worry [about me and the organization].'" General Zinni concluded, "Those are all good qualities, but I found that it honed right in on [competence]."

HOW DOES COMPETENCE AFFECT LEADERSHIP?

Competence is an irreplaceable component of effective leadership. While vitally important, exceptional character and caring for those we lead does not make up for utter incompetence. Conversely, a leader with extraordinary competence may sometimes be forgiven for shortcomings of character and even a lack of care for those they are leading. When the question is life and death, people want to know their leader has the competence to get them out alive. Simply put, people

★★

rarely choose to follow someone who is incompetent. There is a basic expectation that we, as leaders, must know what we are doing. This expectation is not without merit, because our competence has multiple positive effects on our personal success and on the success of the people and organizations we lead. Specifically, as shown in Figure 4.1, competence on the leader's part has been associated with improvements in trust and credibility, adaptability, problem-solving, innovation, and decision-making, all of which lead to increases in success.

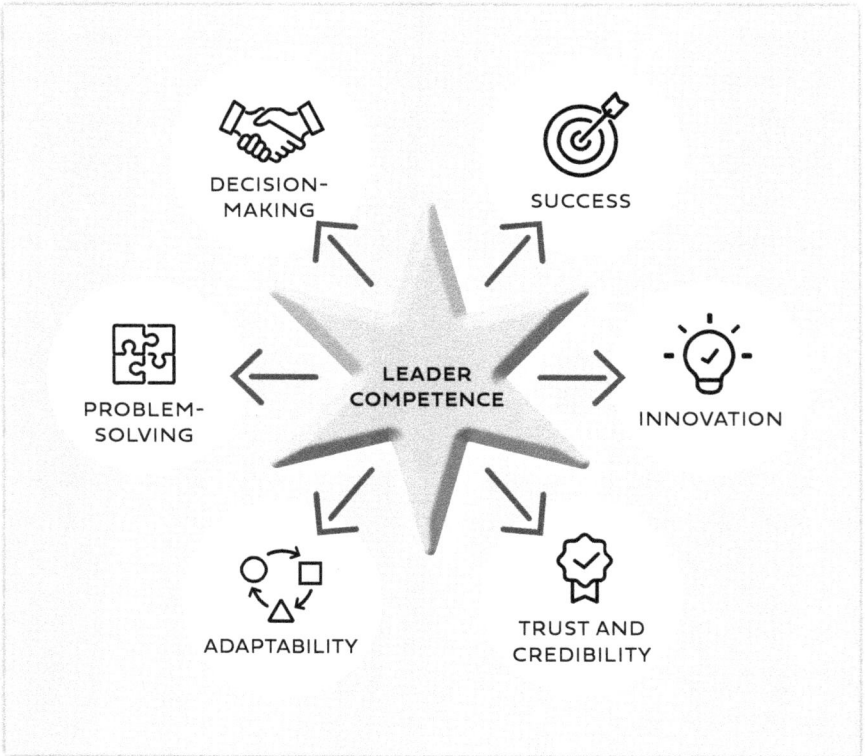

Figure 4.1. The Leadership Effects of Competence. Leader competence results in multiple positive effects for the leader, the led, and the organization. Competence increases the leader's credibility and the trust others place in them. Competent leaders are more adaptable and better at problem-solving and decision-making. These lead to increased innovation in both the leader and the led. All of these effects increase current and long-term success of the leader, the led, and the organization.

★★☆☆

Excellent decision-making

Decision-making is an elemental part of leadership. Both metaphorically and literally, we have to decide where our group is going and how to get there. General Gene Renuart emphasized the essential role of decision-making in leadership as he discussed leadership maxims that have guided him. "A leader has to make decisions. [...] you've got to be able to listen, [...] understand, [...] be compassionate, [and ...] take care of your people, but you've got to also, in the end, make a decision." It stands to reason that if decision-making is a fundamental part of leadership, then excellent decision-making should be a necessary ingredient in exceptional leadership.

When we are competent, we can more accurately analyze complex situations, weigh conflicting data, assess risks, and evaluate potential outcomes. These capabilities put us in much stronger positions to make informed and effective decisions. The ability to make sound decisions is instrumental in Four-Star leadership, as it has been shown to directly impact the success of teams and organizations. General Barry McCaffrey reinforced the relationship between competence and decision-making when he said leaders are expected to "make sensible decisions quickly," adding, "By the way, if you're an expert, you can do that. You listen to everybody talk about the weather, terrain, enemy situation, friendly force [and then make a decision]."

Effective problem-solving

The people we lead expect us to solve problems; like decision-making, it comes with the job. Solving problems well is, then, a part of competent leadership. Competent leaders can identify issues, analyze root causes, and develop effective solutions. General Scott Wallace exemplified this when he said, "The first thing that you need to spend time on, and spend a lot of time on, is trying to figure out what the problem is you're trying to solve. Because most of us, our inclination is to go immediately to the solution. Sometimes it works, but more times than not it doesn't, because you end up solving the wrong [...] problem." General Ed Rice had similar thoughts, saying, "Get the question right. You need to spend enough time to really understand the question [...]

★★

you're trying to solve." Excellence in problem-solving is particularly valuable in rapidly changing environments, where we often encounter new challenges. General Barry McCaffrey linked a leader's expertise to the ability to solve problems rapidly. "I think the key is you've got to be an expert. If you're an expert, you worry about the right things; you don't worry about the wrong things. If you're an expert, you can look at an unfolding crisis and come up with a common sense, simple answer almost immediately."

Enhanced adaptability

Because the world is a dynamic place, if we are to become and remain successful, it is essential that we adapt ourselves and our organizations to the ever-changing landscape. General Steve Lyons equated adaptability with leadership, saying, "The leadership journey is all about adapting and learning and being resilient." As General Lyons suggests, Four-Star leadership means constantly growing, learning, and adapting to the current and future demands of the environment. "You're improving yourself all the time," General David Rodriguez said. "You're growing your capacities." Being adaptable and growing in our capabilities will allow us to navigate change more successfully. Additionally, it can have a direct impact on our career. For instance, General Mike Scaparrotti saw his ability to adapt as instrumental in his success in progressing in his career, saying, "There is no way that I would've made it even to Two-Star or beyond had I not been willing to change some of my habits and learn."

Enriched trust and credibility

Trust is an absolutely essential element of effective leadership. A leader's values and competence have been proven to be important contributors to building trust; that is, followers trust those leaders in whom they don't perceive character weaknesses or incompetence. When we demonstrate expertise and competence in our field, it gives us credibility with those we lead, and they are then more likely to have confidence in our decisions and guidance, all of which result in more effective

★★☆☆

leadership. General Joe Ralston summarized this concept when he talked about being highly competent in one's professional domain:

> Know your profession and be good at it. If you're a fighter pilot, you need to be a very, very good fighter pilot. You need to work hard on your skills, and you need to be out in front leading your people. If you are not up to speed, professionally, that's going to become apparent, and that's going to be a big detractor from your leadership. [...] If you're a division commander of a tank division, you need to know everything about tanks. You need to know tank gunnery. You need to know maneuver. You need to know maintenance. It's very important for the credibility of those people that you're responsible for.

Expanded innovation

When leaders are viewed by those they lead as competent, adaptable, and trustworthy, it fosters innovation in their organizations. General Gus Perna echoed this precept: "You have to be good at what you do to be agile, adaptive, [and] innovative." As leaders, adaptability allows us to view things from new perspectives, and it encourages creativity among those we lead because we are more likely to support and implement novel approaches to problems. Additionally, when we are competent, adaptable, and trustworthy, our people are more willing to take risks with creative ideas because they do not fear negative repercussions. All of these things have been shown to be critical for organizational success.

Elevated long-term success

If we, and consequently the group or organization we lead, are to remain successful over the long-term, it requires enduring competence, which requires ongoing growth. General Mike Scaparrotti said, "Competence is not only knowledge and skill; it's experience, and it's the quest to keep learning." Competent leaders are continuously developing their knowledge, skills, and capabilities, which means they are continuously building the abilities to lead. When we are devoted to continual personal growth, it better positions us and our organizations to handle challenges and the demand for transformation, which

★ ★

contributes directly to long-term success for us, those we lead, and our organizations.

General Paul Kern, *on The Need for Leaders to Build Competence*

Before retiring from the Army, General Paul Kern served as the Commanding General of the U.S. Army Materiel Command and learned a great deal about leadership across his 38-year military career. As he reflected on what it takes to be successful as a leader, he focused on competence:

> Make yourself as competent as you possibly can be. [...] the people that you're going to be working with are going to be looking to you for the guidance and leadership and how to get the things done.

General Kern recalled when he first arrived in Vietnam as a second lieutenant leading a platoon of forty soldiers. The unit was mounted in M113 Armored Cavalry Assault Vehicles (ACAVs), outfitted with multiple machine guns. However, as the platoon leader, then-Second Lieutenant Kern recognized there were several things he did not know when it came to navigating from a vehicle.

> I'd drive down a highway and said, "I've never navigated from a moving vehicle." So, I had to learn how to do that. Asking questions was the first thing I'd do. "How do you use a compass as you're sitting in this little piece of armor around you?" They said, "Stand on the ground, get your compass out, shoot an azimuth and get back in the vehicle and shoot the same azimuth and see what the difference is."

This practical solution from his soldiers helped Second Lieutenant Kern quickly learn how to navigate from the ACAVs. At the same time, he realized there was more he needed to learn when it came to navigating. His soldiers, recognizing he was new to mounted navigation, asked him how he would determine the distance they had traveled. His prior training in determining distance was from the perspective of a foot soldier and not that of someone operating an armored vehicle.

> I was used to walking on the ground and tying knots in a rope every hundred meters [so] I'd know how far I went. They said, "You've got an odometer on the vehicle, just check the mileage on the odometer." I said, "Yeah.

★★☆☆

Okay, that's good." Now, all our maps are in kilometers, and the odometer is in miles. So, I learned to multiply by 0.62 and 1.6 depending upon which way you're going, very rapidly. The competence there was something that I had to learn very quickly and completely differently [than what I was used to]. I did okay with that. I felt confident in doing it, and I could figure out how to shoot artillery from a moving vehicle, and I could shoot. I could maneuver my units in the way I wanted to. I got positive feedback from soldiers because of that. So, that paid off.

This chapter has covered what competence looks like and numerous important effects it has on our ability to lead successfully. The remaining chapters in this section will explore in more depth some of the foundational facets of competence and how various competencies have vital, direct impacts on leadership success. This begins with Chapter 5 addressing the critical need for a leader to "know your stuff," from being an expert, to having vision and a plan, to recognizing the things needed for accomplishing the mission. Without mastering those, you will not be able to lead successfully. Chapter 6 shifts the focus to what is required to maintain Four-Star leadership: continual growth and development. As leaders, we require a growth mindset that says, "What you are today is not good enough for tomorrow."

SELF REFLECTION

Are you as competent as your people and your organization need you to be? What is the evidence to support your conclusion? In what ways might you increase your competence?

★★

CHAPTER 5

KNOW YOUR STUFF

You can't lead unless you're an expert at what you're trying to lead. As you grow up, you need to really be good. Of course, people respect people that are really good [...] In my case it was flying, but it could be anything: could be walking the line, could be grunt, infantry, doesn't matter. If you're not really good at that, then you're going to have a really tough time leading. Even if people like you, it's just not quite there.

– GENERAL CHUCK WALD, U.S. AIR FORCE –

As the Commander of the Eleventh Air Force, then-Lieutenant General Joe Ralston was responsible for the training and proficiency of all the aircrews under his command. To maintain credibility amongst his airmen, he knew he had to know his stuff—what he was doing—as a pilot.

> The way that I did that, and I think the way successful commanders did it, is you have to be out there flying with the troops. When I was Eleventh Air Force Commander, I tried to fly three times a week because that was the only way I could maintain proficiency, and be out there and fly with the lieutenants, and the captains, and the majors, and be helpful in training them. When they did something right, you needed to come back for the debriefing, [and] you needed to tell them what they did right. When they did something wrong, you needed to be able to tell them what they did wrong, and how they needed to change it. [...] Now, in order for that to work, you had to have credibility. You had to be able to go out

★★ ⬩ ⬩

and demonstrate to them that you could fly the airplane, that you knew the tactics, you knew the procedures. Then, they sort of want to be like you, and they will adapt as necessary. They will try to correct whatever mistakes they made. They'll try to reinforce what they did that worked well. [...] In order to be an effective Numbered Air Force Commander, you needed to be out in front, leading the troops where they could see it. They needed to see that you were proficient, and what you told them, in fact, was the right way to do things.

It should go without saying that leaders must know what they are doing. How can you lead someone to accomplish something if you don't know what it takes to accomplish it? People generally recognize this need and expect their leaders to demonstrate a high level of competence in those things in which they are leading. We want our leaders to be experts in what they are doing, because that gives us peace of mind that they know where we should be going, how to get us there, and how to overcome problems along the way. If we, as leaders, do not know our stuff—what we are doing—the people we are leading will not trust us and will not follow us, which destroys any potential we have for effective leadership.

Practically, why would anyone want to follow you if you don't know what you are doing? Doing so would be foolish, and yet we see it happen often as people get caught up in cults of personality, enamored by someone's notoriety. This seems to occur principally in situations where "following" does not carry any immediate and meaningful threat to life or limb. However, even in those situations, the questionable person being followed has accomplished something that certain people want—money, fame, or power—which gives them some credibility as an expert in attaining it. Nevertheless, it is difficult to imagine that anyone sensible would willingly follow someone into a high-risk or life-or-death situation if the person leading is clearly inept. If we expect people to follow us, we must know what we are doing.

As leaders, "knowing our stuff" means being competent at numerous things. As I analyzed the data from the Four-Stars, six major areas emerged in which we must be highly competent if we are to be excellent leaders, and these are presented in Table 5.1 along with the impact

★★ ☆ ☆

they have on those we lead. The remainder of the chapter will cover these facets of "knowing your stuff."

Table 5.1. Major Areas in Which Leaders Must Be Highly Competent

LEADER COMPETENCE	IMPACT ON TEAM
1. Expertise	Credibility for the leader and trust among those we lead
2. Clarity of vision	Provides direction that produces unity of effort
3. Ability to anticipate the future	Allows for the development of flexible plans that shape the future
4. Ability to create the environment for success	Removes obstacles and sets up the team to be able to focus on the mission
5. Strong planning skills	Produces understanding of the environment, resources, and actions needed for success; creates the framework to allow flexibility and change as the environment changes
6. Ability to accomplish the mission	Increases team confidence in the leader and themselves

HOW DO YOU BECOME AN EXPERT?

Long before General Keith Alexander was the Commander of U.S. Cyber Command and the Director of the National Security Agency, he was a young Army intelligence officer. As a captain, he had just stood up a new company, and the Army was planning to deploy a highly technical intelligence system, but few people knew anything about it. Captain Alexander was assigned to learn the system.

There was a 12-week training program out on the West Coast, learning the math behind the program. It was exceptionally hard and a novel approach

★★ ★ ★

for the military and for intel for what the XVIII Airborne Corps needed. So, [...] I'd go out every other week. I'd sit down with the technical lead for it and say, "Can you show me how the predictor-corrector algorithms work? I just want to understand it. Help me understand the math, help me understand where this goes. How does this work? How does [that] work?" So, I learned over that 12-week period how to actually write out the design of that whole system on a board without any other paper. I knew it, I memorized it.

Through the training program, Captain Alexander developed a high degree of expertise in the new system. When he returned to his base, he was asked to give successive briefings on the new system to his supervisor, his supervisor's boss, and, shortly thereafter, the brigade commander.

Now the new brigade commander comes in [...] and says, "Can you give me an operational lay down of how this system works?" I thought he meant technically. So, I got up on the board, and I got a piece of paper, and I said, "So here's how it works." I laid out satellites. I laid out the theory of the beams, how the predictor-corrector algorithms work, and what this meant for [the system]. It went for about 45 minutes. That guy was chewing a cigar, just watching it and just looked at that, and he said, "Really good." He got up and left, and the deputy comes walking in a few minutes later. He [says], "I don't know what you told him, but he's hugely impressed. But he just wanted to know how you use it operationally for the military."

Because Captain Alexander had developed and demonstrated superior expertise in the new intelligence system, it had far-reaching repercussions for his leadership as he moved through his accomplished military career. "I think being an expert is the key part," he said, "and it helped for deploying that system, for the things I did at the Intel Center and School, for the things I did on the Army staff, all the way up for the things I did running Cyber Command."

One after another, the Four-Stars said if you are going to be the best possible leader, it's critical to be an expert in your field. For instance, General Frank McKenzie said, "You have to be a master of your craft," and General J.D. Thurman echoed that perspective saying, "You've got to know your business. You've got to be the best at that." Others, such as

★★

General Paul Kern, included the commitment to personal development of expertise as a key component, saying, you have to "make yourself as competent as you possibly can be," something General Bob Kehler did by "be[ing] a student of the profession." Perhaps none of the Four-Stars was more adamant about the requirement for expertise than General Chuck Wald, whose quote opened this chapter. He effectively synthesized the perspective of most of the other Four-Stars regarding why expertise is important when he said, "You better be pretty damn good at what you're doing, because you have no credibility telling other people how to do things if they're better than you."

While expertise in a particular area is crucial, it is not humanly possible to be an expert in all things required in any large, multifaceted organization. No one could expect that from a leader. Instead, what is expected is expertise within that leader's particular field of training and practice. For fighter pilots, the leader must be an expert fighter pilot. The same goes for leaders in myriad domains, from computer science to accounting to nursing to journalism. When we have expertise in our professional field, we have credibility with those we lead and can more directly and correctly speak to what is happening. When a leader has no expertise, none of that can happen.

Beyond domain-specific expertise, achieving Four-Star leadership requires we have the leadership expertise to work well with people and direct our group or organization—things covered in the remainder of this book. For many, it is the need to work well with others and direct the group that prevent them from being excellent leaders. This phenomenon is described by the Peter principle, the idea that people with great expertise in one area will eventually get promoted to the point where the demands of the role exceed their abilities, what is often referred to as the point of their incompetence. Many times, that threshold of incompetence is a position of leadership where their domain-specific expertise isn't what is required in their new role.

★★ ☆ ☆

WHAT DOES IT LOOK LIKE TO HAVE A CLEAR VISION?

Prior to becoming the first female General in the U.S. Air Force, Janet Wolfenbarger spent decades of her career serving in roles in aircraft acquisition—the design, development, and delivery of new aircraft for the Air Force. Those varied and challenging experiences helped her develop the extraordinary expertise to revolutionize the organization of Air Materiel Command and save $6 billion (see the beginning of this chapter).

> [My expertise at streamlining] came from my jobs leading up to that Four-Star job. So, I was put in charge of some of the acquisition programs in the Air Force. [I] spent eight years on the F-22 program. [I] wasn't in charge of that one early in my career, but I learned a ton about how you go about Air Force acquisition. [I] was put in charge of the B-2 program and [...] the C-17 program. The way Air Force acquisition works, [...] we get graded by the system on a stoplight chart, red, yellow, green, on the various aspects of that program—cost, schedule, and technical performance are the key ones. So, you establish a baseline [at the start of the program] from which you're going to be judged.

As she led the various programs to develop new aircraft, General Wolfenbarger and those she led had to repeatedly reorganize and reorient due to external demands.

> [I was] held accountable for that stoplight chart despite the fact that much of what affected it was outside of my control. Congress would cut my budget. That would drive a replan. The replan would result in not being able to achieve the originally established goals for cost, schedule, and technical [specifications]. The user, the warfighters who established our requirements for each one of these platforms, sometimes [would] change their mind [or realize they needed more capability in the aircraft]. [...] That drove, again, replans.

Through the constant demands for changes to budget, schedule, and technical performance of the aircraft, General Wolfenbarger developed expertise in creating processes "to ensure the highest probability of success in delivering this warfighting capability despite those disturbances."

★★

There really were no standard processes in each one of those program offices that I worked in. Everybody did it differently, and the thought had occurred to me early in my career, if I knew a process that had succeeded in the past using the same kind of environmental constraints that I'm in now, then I would love to lift that and apply it in these programs that I'm in charge of.

It was really my learning and the understanding that there were areas that were succeeding better than others in Air Force acquisition, and [then asking the question] "Why is it that we couldn't discern what was allowing that and then levy it as a standard process [...]?" We were missing this huge opportunity to get better at this really important work.

General Wolfenbarger took these insights and her expertise into leading Air Materiel Command, which was faced with a tense situation demanding substantial budget cuts, creating a high degree of uncertainty throughout the organization. Drawing on all her prior experience, General Wolfenbarger created a vision for how the organization could be streamlined, and then she set about communicating that vision clearly to the organization through her strategic plan. She noted that in her prior experience with strategic plans they often took a tremendous amount of effort to compose, and then people put them on the shelf never to be read or used.

What I realized was I couldn't do it that way. I had to establish for my command [...] the vision [...] for this approach that we were taking, and I needed to establish a strategic plan that would be leveraged by every single person in my command. So, I made it short. It was like 16 pages long. I had my commander priorities identified in it—five of them. I made sure that every single one of my airmen saw themselves at least in one of those five commander priorities, many of them more than one. Then, we put in place these metrics that we started tracking...

Soon, personnel throughout Air Materiel Command had received General Wolfenbarger's clearly articulated vision, and the massive reorganization was instituted with extraordinary success. General Wolfenbarger determined that having a clear and concise strategic plan—the vision—is "essential [...] but they have to be readable [...] digestible [...] actionable, and you have to use them."

★★☆☆

To be able to lead anyone, you must know where you are going. You must know what the goal is that you are trying to accomplish. What is the ideal end-state toward which the group or organization is working? This is your vision for those you are leading. If you don't know where you are going, people are unlikely to follow you. As a leader, you must be able to visualize the goal and how to make it happen within the limitations of personnel and resources. Beyond that, you must be able to communicate your vision to the people you lead (covered further in Section Four on Communication). The ability to visualize the goal and how to make it happen, and then communicate it clearly to those we lead is, in a real sense, the essence of excellent leadership.

WHAT DOES IT MEAN TO INVENT THE FUTURE?

General Tony Zinni served as the Commander of U.S. Central Command (CENTCOM) from 1997 to 2000. In General Zinni's recollection, at that time CENTCOM was largely "focused on Iraq and Iran," though they also had responsibility for the remainder of Central Asia, including Pakistan and Afghanistan.

> I was sort of watching Afghanistan, because we were trying to get at Al-Qaeda, and the Saudis were making all sorts of overtures to the Taliban, offering them money and everything else to kick out Osama bin Laden, so they could grab him or turn him over, and it always looked like they were making progress. [...] The Taliban always looked like they were on the verge of cooperating [in] some way. So, I didn't get overly focused on that.

In the meantime, Al-Qaeda continued to organize and were becoming more regionally active. Then, on August 7, 1998, Al-Qaeda suicide bombers simultaneously detonated trucks loaded with explosives at the U.S. embassies in Nairobi, Kenya and Dar es Salaam, Tanzania, killing more than 220 people and wounding over 4,000. Fifty-six of those killed were U.S. government employees, contractors, and family members. Nairobi was in CENTCOM's global area of responsibility, and they had not anticipated the Al-Qaeda attack.

★ ★

We were suddenly looking for targets, and I realized we had all these war plans for Iran, Iraq, [and] everything else, [but] we had not really thought through Afghanistan. [I began to ask myself,] "What if we had to go in there?" "What if we had to take out the Taliban?" "What if we had to go get this guy and have to do it with overwhelming force?" [...] So, I would say the biggest lesson I learned from all that is you have got to look more broadly at all the things that can bite you, because there's something out there that you're not thinking a hell of a lot about or maybe not thinking about at all that's going to come up and bite you.

The best leaders can look at the current environment and see where things will be moving. Admiral Pat Walsh said leaders must be "always out there looking for data, for facts, and at the same time [must be] able to see where the ball or the puck is moving..." Four-Star leaders look at the terrain, whether literally or figuratively, and can anticipate how things are going to unfold. In military history, this has sometimes been referred to as *coup d'oeil*—the ability to discern immediately the lay of the land and how the battle will unfold. "Anticipation is the key to leadership," according to Admiral Walsh, and its importance has been demonstrated in numerous fields, whether through successes, such as Intel's storied shift from memory to microprocessors or Apple's anticipation of the need for personal computers, or failures like Kodak's refusal to embrace digital photography or the multitude of people who did not anticipate the devastation a direct hurricane strike could have on New Orleans, LA. For people and organizations, leaders' ability to anticipate the future can mean the difference between life and death, for people and organizations.

Predicting the future is nearly impossible, but anticipating the future and being prepared for its potentialities is not only possible but likely, given the right circumstances. General Bob Kehler said instead of waiting for the future to happen, the best leaders "have a strategy that invents the future." He continued, "If you're waiting for a future to unfold the way some expert says the future's going to unfold, it's never going to happen that way." How then can we anticipate and, to whatever degree possible, "invent" the future? Admiral Tom Collins developed a mental process to help him anticipate the future.

★★ ☆ ☆

I always tried to prepare the battlefield physically and mentally. I always tried to think ahead. I always tried to think of what could happen. Could we run into sloppy weather? [...] those kind of things. [...] that's a part of good leadership [...], knowing the environment, preparing mentally and physically, and preparing a battlefield.

General Glenn Walters expressed this concept concisely: "Know your battlespace, anticipate challenges, and prepare responses for these things." As Figure 5.1 shows, leaders can shape the future by using a cyclical process.

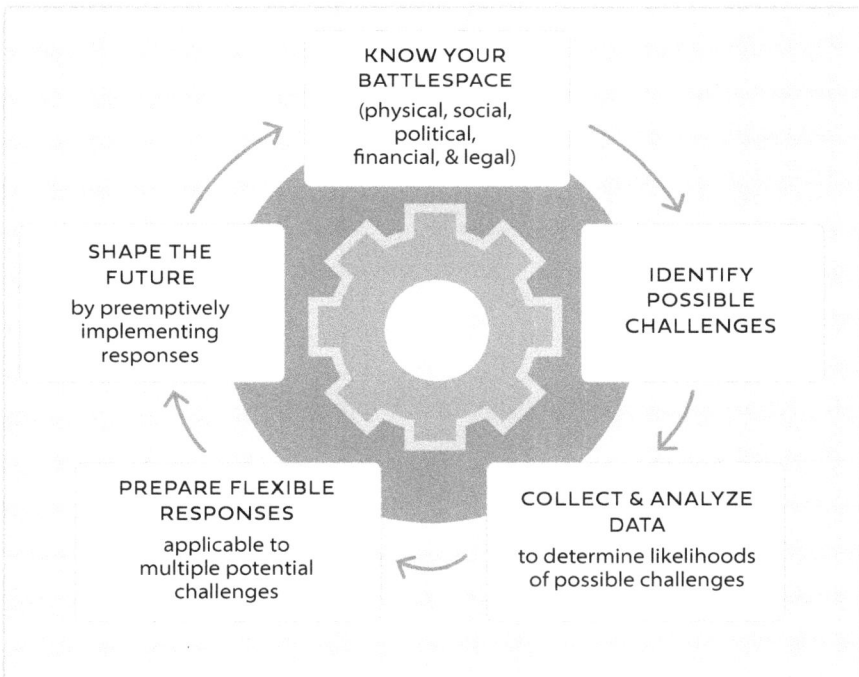

Figure 5.1. Anticipating and Shaping the Future. To be able to anticipate and shape the future, leaders must study and understand their environment—their battlespace— including the physical, social, political, financial, and legal contributors. Leaders then must identify possible challenges they might face and collect and analyze data to determine how likely it is any of those challenges could occur. Leaders then must prepare flexible responses that allow for application across multiple possible outcomes. By preemptively implementing one or more of those responses, leaders can shape or create the future, which will change the environment and necessitate reinitiation in the cycle.

★★

HOW DO LEADERS CREATE THE ENVIRONMENT FOR SUCCESS?

Former Secretary of Defense Donald Rumsfeld was notorious for being a hard-charging, demanding person who was hard to work with. He didn't appreciate feedback or being questioned. So, when General Gene Renuart became his senior military assistant, General Renuart knew to be successful he had to create a way to be able to ask questions of Rumsfeld without upsetting him.

> When I started with Secretary Rumsfeld, I said, "Mr. Secretary, I need you to allow me at least one 'but sir' a day." He [replied], "What do you mean 'but sir'?" I said, "Well, you're going to see and hear a lot of briefings and have discussions, and you'll give out guidance, and all of that's critical." I said, "But my bet is that probably one or two of those folks won't completely understand what you were driving at or the guidance you gave, or we'll have questions. We don't have time in the office time that you have to answer all those questions. So, I need to be able to come back at the end of the day and say, "But sir, you said you wanted this, this, and this done. Could you clarify for me, and then I'll go help whoever that team was to try to get them going down the right road?"

Secretary Rumsfeld brusquely agreed to General Renuart's proposal, allowing him to return to his office at the end of each day where Rumsfeld would ask, "Okay, what's your 'but sir' today?"

> It created for me a safe place [where] I [was] not question[ing] what he said, but rather [where I could] clarify in my mind, and hopefully, maybe clarify in his mind, the guidance he gave out in some form through the course of the day that might not have been taken exactly the way he meant it. So, the ability to go back and clarify with your boss, because we all have one, is really important because it's the chance to ask that question.

At the time, when the U.S. was waging wars in Afghanistan and Iraq, the clarity of communication in the directives coming from the Secretary of Defense was imperative for success. General Renuart's use of the "but sir" approach created the environment for him to be successful in his role. More importantly, it allowed him to achieve clarity for himself, Secretary Rumsfeld, and anyone else to whom the Secretary's

★★☆☆

instructions might be directed, which effectively meant the entirety of the U.S. military.

Success does not just happen, either on the individual or group level; it requires preparation. In the case of leadership, preparation goes far beyond making certain those we lead are trained and equipped to accomplish the mission. If we want to be best prepared to be successful, we have to create the environment for success. Nearly 70% of the Four-Stars discussed that. For instance, General Mike Scaparrotti said, "A proper command environment is established by the leader and is marked by trust, teamwork, inclusivity, and respect. [...] I think the environment is a reflection of the leader, and it's the most important thing in an organization." Creating that proper environment directly affects the likelihood of your group's success—the more you're prepared, the higher the likelihood of mission success.

As alluded to by General Scaparrotti, an environment conducive to success includes multiple factors. Many of those factors are covered throughout the course of this book; however, I have identified four major actions leaders can take to create the environment for success, shown in Table 5.2. What do each of those look like for the Four-Star leader?

Table 5.2. The Major Actions of Leaders Who Create an Environment for Success

LEADER ACTION
1   **Do the basics exceptionally well**
2   **Create the culture**
3   **Harness the power of diversity and inclusion**
4   **Prepare for success *and* failure**

★★

Do the basics exceptionally well

In June 1986, then-Lieutenant Colonel Pete Pace was assigned as the Chief of Ground Forces in the Combined Staff in Seoul, South Korea. There was a two-month lag time from his arrival until his family would join him, and Lieutenant Colonel Pace used that time to study for his new role.

> I was really concerned about the war plans on the peninsula and my adequacy. [...] I felt like I needed to know my job, and I didn't know it well enough. So, there were two filing cabinets, each with four drawers in them [...] eight [drawers...] of previous notes, war plans, et cetera. I would do my job during the day, and then I would stay at night, and I would just read. I read everything in those filing cabinets—everything. I was nervous, [thinking to myself,] "I need more data. I need more information. I need to be smarter than I am."

After spending long hours into the nights poring over the war plans, Lieutenant Colonel Pace was well-versed in the military plans for the Korean Peninsula—the basic operational plans for how to protect and defend South Korea from invasion. This had an immediate impact on his success and the success of the unit.

> I started going to meetings. People would start saying things that were not true; not that they were lying, but that they hadn't read the plans. I would say, "No, wait. That's not what the plan says. The plan says this." So, I got to be known very quickly as the go-to genius who knew all this stuff about the war plans, and all I had done was read them.

John Wooden was an exceptionally successful basketball coach for the UCLA Bruins, leading his teams to ten national championships in twelve years, including seven in a row. He was dedicated to teaching his players to do the basic exceptionally well, so much so that at the beginning of every season, he would teach them how to put on their socks correctly so they wouldn't get blisters. If the foundation is weak, you can never expect to build something great, whether literal or figurative.

Competence—mastery in any given domain—requires first having mastered the basics. Leadership is no different. You simply cannot be

★★☆☆

a good, let alone great, leader if you have poor character, are incompetent, don't care about those you lead, or cannot communicate with them. General Mike Scaparrotti said, "Regardless of rank, a leader must know and practice the fundamentals." He went on to say, "High performance units do the fundamental tasks routinely and to high standards." Those units follow the example of their leader. Additionally, General Mike Murray identified a particularly important reason that doing the basics well is vital to organizational progress and success.

> Organizational energy is finite. So, if you take all the energy trying to do routine things or focus on routine things, you'll never get to that next level. So, it's always, to me, been about basic blocking and tackling, doing routine things routinely well, so that you can get to the next level.

As covered in a prior chapter, those we lead are always watching to see if we know what we are doing, if we can be trusted to lead, if we are worth following, and if we are safe to follow. When people are evaluating a leader, they don't necessarily look specifically at the basics, because they assume those are there. But, when the basics are absent or done poorly, they force themselves into the spotlight. If you can't do the basics, how can you expect to do the more complex things as a leader? A leader must be able to do the basics well. Notably, being able to do the basics well does not assure leadership success, but an inability to do them well essentially guarantees leadership failure.

Create the culture

As a young captain, James Conway became Commander of Kilo Company, 3rd Battalion, 2nd Marines, replacing a commander known to be "not a very strong leader." At that time, Kilo Company was in poor shape with regard to preparedness, morale, and *esprit de corps*. General Conway recalled, "I had a lot of fallow ground there that I could really turn into something positive..." He told his Marines, "We're going to work hard, we're going to play hard, and [...] if those first two things work out well, I'm going to give you some time back..." To begin addressing those things, Captain Conway started with "one of the banes of a company commander"—physical training (PT). Every day, numerous Marines would find excuses to miss PT. Because their absence eroded

★★ ☆ ☆

both preparedness and *esprit de corps*, he made it a priority for every-one in the company, requiring those who missed it to make it up on their own time.

> We were working towards a run from our barracks out to the main gate at Camp Lejeune, which was a 13-and-a-half-mile trip. That's half a mar-athon. Now, I'm not a runner, okay. I weighed 205 pounds. I was great in the 440, but boy, I hate distance. [...] But it was really good for us. It helped build morale, and my straggler numbers went from 10 or 15 for each PT session to zero because, all of a sudden, they're doing the same thing [they] could have done that morning, [but now it's] on their time, and nobody liked that.

Captain Conway demonstrated himself to be reliable and trustworthy to his Marines. He made sure they worked hard to get into shape and build morale, and he also lived up to his end of the bargain and re-warded them with time off when they did so. "On Fridays, if we had had a good training week, I would call the company together for our formation at 13:00," General Conway recalled, and he would release his Marines "early into the rest of the weekend."

After some time getting Kilo Company turned around, they went into the field to conduct a training exercise in a head-to-head against India Company, with the objective of patrolling and then conducting a night attack against each other. India Company conducted their night attack first, but it failed. When it was Kilo Company's turn to go on the offensive, General Conway recounted, "We moved out in the wee hours. We hit them [...] between 3:30 and 4:00, and it was a walkover. [India Company's] sentries put up some feeble resistance," he said, "but we were in the middle of the camp almost before we knew it. The troops were [...] high-fiving. They were excited as hell [...] that kind of thing just made the morale zoom."

Later, all the companies in 3rd Battalion competed in a field meet consisting of an assortment of events. With Kilo Company having been through the transformation under Captain Conway, their preparedness, morale, and *esprit de corps* were soaring as they went into it. With great pride and fondness, General Conway reminisced on Kilo Company's performance. "We scored 97 points. The next closest company [scored]

★★☆☆

26 [points]. So, that environment just meant so much. Being able to craft that sort of morale was fantastic and probably the most fun I've ever had as a commander. [...] If you weren't participating down on the field, you were up in the stands [chanting], 'Kilo! Kilo!'" Relishing that memory, General Conway said, "The staff walked away saying, 'Holy cow. We knew who our best company was. We didn't realize it was so lopsided.'"

SELF REFLECTION

How much effort are you putting into building the culture of the group you lead? What would those you lead say about the organizational culture?

Throughout the leadership literature, the idea of organizational culture has garnered a lot of attention as being a significant contributor to success. What is really meant when people talk about culture from a leadership perspective? Anthropologically, a culture represents a group's collective beliefs, traditions, language, practices, social structure, music, art, and more—the large set of things that sets them apart as a distinct entity. Culture also allows individuals to identify as being a part of the group because of their participation in those shared things. While some of these elements contribute to the culture of a given organization, General Steve Lyons suggested organizational success "comes from a culture [...] of dignity and respect where everybody feels as though they are empowered to contribute to their maximum potential." So, from a leadership perspective, creating a successful culture is much more about building relationships and camaraderie among the members of the group. To begin to build those sorts of relationships, leaders have to ask ourselves certain questions: "How do we view each other? How do we treat each other? How do we work together? How do we go about doing those things we are supposed to be doing?" We can

★★

then begin creating a winning culture. But, the question arises, how does a leader create such a culture?

General Conway's example of leading Kilo Company provides a masterful demonstration of how creating the right culture can transform a broken and poorly performing group into a highly successful one. Additionally, it outlines some key lessons for how to go about creating camaraderie, teamwork, and *esprit de corps*—the culture—needed for success. As shown in Figure 5.2, General Conway did four major things to change the culture of Kilo Company, and this set up a cycle for increasingly improved *esprit de corps* and culture. First, he **Engaged** with his Marines, establishing what he expected of them and what they could expect of him. At the same time, he showed them he was one of them by knowing what was important to the group. The combination of follow through on expectations and being one of them created trust, a vital component for strong culture and leadership success. General Conway provided **Motivation** to his Marines in two clear ways. He established a reward (time off) that demonstrated he understood what was important to them, and he actively participated with his Marines in various training, whether long runs, nighttime training raids, or otherwise. These activities further created trust, as well as camaraderie. General Conway trained his Marines to perform their jobs well and as a team. Finally, due to the hard work and training they had received, the Marines of Kilo Company trounced India Company in their field exercise, which gave them all a deep sense of pride in themselves and their Company, as well as a great **Sense of Accomplishment**. Thereafter, because of the transformation of the culture of Kilo Company, they went on to dominate the field meet and were recognized as the best Company in the 3rd Battalion, 2nd Marines.

Harness the power of everyone's perspective and input

When General Ed Eberhart's forces were preparing for an inspection by the Inspector General, they discovered a major issue with their aircraft's defense systems. The aircraft had electronic countermeasure pods that jammed enemy radar systems and prevented the planes from being shot down by radar-guided or infrared missiles, and they

★★☆☆

GROUP PRIDE

Provide opportunities for group accomplishment

ENGAGEMENT

- Meet your people
- Build rapport
- Provide expectations

WINNING CULTURE & ESPRIT DE CORPS

TRAINING & EXPERTISE

- Train individuals and teams
- Aim for exceptional, not mediocre

MOTIVATION

- Reward them
- Participate in activities with them

Figure 5.2. Creating a Winning Organizational Culture. Leaders can build a successful organizational culture by engaging directly with those they lead, motivating them to work toward the mission and with each other, providing training of individuals and teams with the goal of being exceptional, and providing them opportunities to work together to accomplish things and develop group pride.

were having trouble reprogramming the countermeasure pods in the allotted time. Given the nature of the problem, this meant not only would they fail the inspection, but also that if they were activated for a mission, it would pose a very real threat to the pilots flying that mission. Leaders across the base had been trying to solve the problem for several days, but they were making no headway because the colonel in charge of maintenance was certain they were doing it as fast as humanly possible. General Eberhart realized to solve the problem he needed to get some ideas from those with a different perspective.

[I] went down to where the young airmen were who were reprogramming these pods and said, "Hey, look guys, how come we can't get it done during this period of time that you have to get it done to pass? How can

★ ★

we do this?" One airman raises his hand and says, "Sir, if we did this, this, this, this [...] then, we'll get it done, but that's not how our local procedures are right now." I [said], "Duh." I should have gone down and [seen] the airmen two days earlier. That's being inclusive. He knew how to do it. You know what I had asked him? I said, "Why didn't you say something?" I didn't say it in an accusing [way], I'm just [thinking], "God, you knew this [...] How come you didn't say something?" What he said is very telling. "No one ever asked me. No one ever asked me."

No two people have exactly the same life experiences, perspectives, and abilities. Everyone has something to offer, and in any given circumstance, that something may be the exact thing that makes the difference between success and failure. In a real way, those varied things our people have to offer are resources—vital resources—that we can, should, and must harness for accomplishing the mission. What leader in their right mind would actively avoid using all available resources to achieve the goal? Sadly, probably more than you realize. In fact, in a 2019 report by Nate Dvorak and Ryan Pendell from the Gallup organization, only 30% of U.S. employees felt their opinions were listened to or counted. That's an enormous, missed opportunity for getting better insight, ideas, and innovation.

Often, leaders have gotten to their positions because they have demonstrated capabilities at lower levels. When we have been successful as individuals and have been rewarded for it, it can be easy in the new role to think we know what we are doing and to continue operating as we have previously, believing we have the knowledge and answers needed to solve the problems we face. As a result, we may ignore the inputs of others, especially if they are at a lower organizational position than we are. Additionally, we may even feel their ideas are an attack on our thoughts and abilities. This bias toward our own perspective limits our ability to have a broader view of things and decreases the options we have available to address challenges. General Dick Myers identified this as a problem for some leaders and the need to be open to others' ideas if we are to be successful.

You've got to be open to everybody's ideas. It's a complicated world today. [...] It's really complex. So, none of us have all the magic answers.

★★☆☆

But when you get a bunch of people around a table and if they feel em-powered, [...] they're going to tell you when they think you're wrong. You might not like that, but you put all the ideas together, stir them all around, and out of that will come probably a pretty good decision. If you try to do it on your own, you're not going to last very long, frankly. [...] It's a hard world. It's a more complex world than it was three or four decades ago. I think you've got to be really open to new ideas, even some that you might not agree with. You've got to be comfortable that other people's ideas matter, and that's not putting you down.

By giving everyone an opportunity to contribute their perspectives and ideas around particular challenges, you can greatly increase your options for ways to overcome those challenges. The more options you have, the more likely you are to have multiple options for success. This is fundamentally the reason the Four-Stars were strong proponents for diversity and inclusion—diversity of thought and experiences increas-es group capabilities and impacts effectiveness. As the first female Four-Star officer in the U.S. Military, General Ann Dunwoody discussed her desire to select the best people from all walks of life and how that diversity strengthens organizations.

Leverage the power of diversity. [...] a lot of people [...] believe diversity [is] about numbers, "One of these, one of those," instead of leveraging the best and brightest from all walks of life, and it's not about numbers. It's about [...] the talent out there, and there's so much. [...] Without [having a diverse group helping you make decisions], you're not going to get the input from the best and brightest to help you solve these complicated problems.

This idea of harnessing the ideas of everyone resonated with Gener-al David Petraeus, who said if leaders are going to "get the big ideas right, it's best done collectively. It's inclusive; it's iterative; it's everyone engaged."

People must be given a voice if we are going to be able to capi-talize on all capabilities and capacities available to us. This does not mean everyone's ideas or opinions are valid for any given circum-stance, but we will never know that if we don't give them a chance. Additionally, it is usually those closest to the problem—the ones on

★★

the frontlines—who have the most up-to-date and informed perspective on a given challenge, which is what is needed to overcome those challenges, especially in rapidly changing environments. Rarely do those on the ground need a leader to tell them how to overcome a given challenge; instead, they need the leader to ensure the resources are available to do what needs to be done to overcome it. When leaders fail to involve their team in shaping a plan to address a challenge, they disregard the input and expertise of those responsible for executing it. This mistake guarantees the plan will be made based on old information and by those who do not have the closest perspective. In rapidly changing environments, that introduces a great deal of risk that can be avoided by giving everyone an opportunity to speak.

Prepare for success and failure

In October of 1983, Michael Rogers was a Navy lieutenant junior grade in charge of the team that fired the large guns on a destroyer. In preparation for their upcoming tour of the Mediterranean Sea, Lieutenant Rogers had led the team through gunfire training on the range and received the necessary certification that they were combat ready. Soon after steaming out of Norfolk enroute to the Mediterranean, the plan changed. He recalled, "Literally, within a day of leaving Norfolk, we get the classic naval message, 'Detach from formation and proceed at best possible speed to...' [particular coordinates]." Everyone onboard was confused as to why they were being sent to that location. Admiral Rogers remembered, "We're looking at it going, 'What? [...] The only thing that's close is this little island, Grenada. Who's ever heard of this place?'"

After the destroyer arrived off the coast of Grenada, a Navy SEAL team was sent onto the island, and then-Lieutenant Rogers and his team stood by to provide naval gunfire support. Admiral Rogers recalled the intense situation.

> We're getting ready to do naval gunfire support for a SEAL element that is in contact at a radio station north of the capitol. They're pinned down by some Cuban troops, and they're [...] trying to break contact so they can get their wounded out, and they're looking for us to just do some

★★☆☆

naval gunfire so the Cubans [would] put their heads down, [allowing the SEALs to] break contact. So, lots of tension. [...] We're running our first fire mission ever. This is combat, lives on the line. I've got a live target. I've got SEALs on the radio going, "Okay, here's the grid."

Lieutenant Rogers and his team were preparing to fire their guns, but there was a sudden, inexplicable delay in communication. Admiral Rogers recalled, "I am waiting [and] waiting... The skipper's right next to me, and he's going, 'What's going on? What's going on? Why the delay?'" Lieutenant Rogers was waiting for word from the ship's bridge, but it was not coming. Rogers turned to one of his sailors who was in communication with the bridge, "What is going on with the bridge?" he asked. The sailor responded, "'Sir, it sounds like utter [bedlam]. I hear people yelling and screaming up there.'" Recognizing that time was of the essence, Lieutenant Rogers sprang into action and ran up the ladder to the bridge. where he found:

> It is utter chaos [...] because one member of my team who was taking his visual lines of bearing [had] suddenly decided that he cannot take a life. And I'm thinking to myself, "Okay. We are in a combat situation. We've got troops calling for fire. We've got troops in contact. We've got U.S. SEALs wounded. They're trying to break contact, saying get their guys out, and you've decided now you can't do anything?!"

Lieutenant Rogers immediately relieved the sailor from his post, tasked another petty officer to sight the visual solution, and they were able to fire the mission.

Looking back on the experience off Grenada, Admiral Rogers saw it as being "the most impactful thing [he] ever did in 37 years" of military service, one that continues to shape his leadership.

> I failed as a leader. I thought to myself, "Michael, you didn't sit down and talk to your team. You just treated this as [if] it's just another [training] evolution." [...] I never sat down with the team and said, "Okay, let's walk through what this really means, and if we were going to do this that we potentially are going to take human life and harm and injure and destroy, and is everybody comfortable?" I kicked myself, "Rogers, basic failure as a leader."

★ ★

Much of the leadership literature is focused on how to prepare for and achieve success, but there is little attention on how to prepare for failure. That can leave us and those we lead in a vulnerable position. Preparing for failure may be as important to accomplishing our goals as preparing for success. General Tony Zinni said, "You never fight the wars you prepare for, and the reason you don't is because you prepared for them." If we prepare for failures, we will do things to prevent them from occurring. It had not occurred to then-Lieutenant Rogers that one of his team would refuse to do their job in the heat of combat operations; had he recognized that potential and prepared for it, it wouldn't have occurred. To create the environment for success, we have to prepare for failure.

The idea of preparing for failure has gained some traction recently in the form of the "pre-mortem examination." In this process, a group imagines a project has failed and examines the causes as if reflecting back on the failure. They assess the likelihood and impact of each cause to identify the highest risks, then develop solutions to mitigate them. I have synthesized this concept into Figure 5.3, that provides a general approach to threats identified during a pre-mortem based on their degree of likelihood and impact. For those threats that have a low likelihood and low impact, it is best to **Accept** them and budget accordingly. For those that have either a high likelihood and low impact or a low likelihood with a high impact, it is best to try to **Mitigate** them by acting to reduce the likelihood and/or the risk. Finally, **Address** those that pose both a high risk and high likelihood by doing everything possible to attack the drivers of those threats while simultaneously creating defenses against them.

While preparing for failure is important to success, General George Casey admonished that leaders must keep an important perspective on preparation and planning. "There's only two kinds of plans," he said, "those that might work and those that won't work, and the best you're going to do is a plan that might work." He continued, "I've found [...] if you're humble enough to accept that, then you can still reach high and stretch yourself. But you don't flog yourself unmercifully if you don't quite get as high as you wanted."

★★☆☆

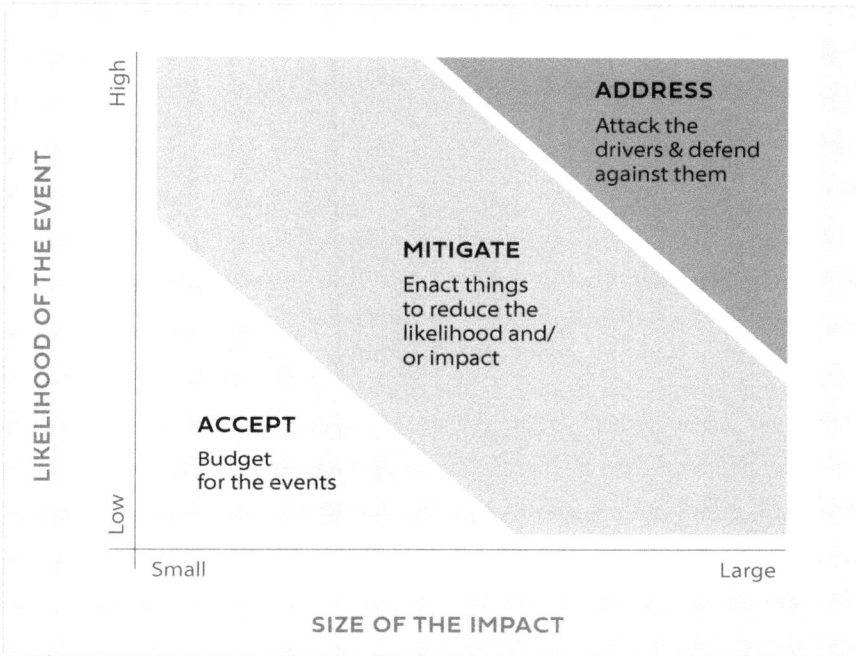

Figure 5.3. Approach to Threats Identified During a Pre-Mortem. Those threats with a high likelihood of occurrence and a high impact must be addressed. Those that have a low likelihood of occurrence and a small impact should be accepted, budgeted for, and disregarded. The remaining threats must be mitigated by enacting things to reduce the likelihood and/or magnitude of the impact.

WHAT IS THE VALUE IN HAVING A PLAN?

On February 24, 1991, U.S.-led coalition forces raced north out of Saudi Arabia into Iraq before turning sharply eastward in a move that General Norman Schwarzkopf referred to as the "left hook." The massive ground attack drove Iraqi forces to either surrender where they were or make a hasty retreat out of Kuwait, ending the fighting 100 hours after the ground assault had begun and establishing it as one of the most successful victories in U.S. military history. However, prior to that, then-Major General Barry McCaffrey, the Commander of the 24th Infantry Division (Mechanized) that were heavily involved in the "left hook," thought the battle would be anything other than swift and decisive. "We thought it was going to be World War III," General McCaffrey said.

★★

"Almost everyone in the intelligence superstructure said this was going to be worse than the Normandy invasion." The concern over how the attack could play out led to thorough planning and preparation prior to the attack.

> I said, "We deserve to treat this with seriousness." So, my division had a plan [...] we all came together, and we started a series of, essentially, sand-table exercises [...] maybe four or five of them in Saudi Arabia before the attack started. We had a tremendous sand table, 20 feet by 20 feet. We'd walk through each, and then we'd talk about it. Then we'd start the next phase [...] and it would go on until midnight, and I'd let them sleep for four hours. But, sometimes they'd be two days long. So, it wasn't just [that] we had a plan, which we did, but we'd rehearsed it innumerable times, and everybody had heard what everyone else was going to do.

The intensive preparation for the attack gave his commanders a great depth of understanding of the plan. They knew everyone's role in it to a degree that allowed them to quickly adjust as things on the battlefield evolved. By knowing the plan intimately, they were able to make changes in real-time without negatively impacting the other commanders' operations.

> Then we started the attack and [the plan] had to change, but it changed from a common understanding and grasp of every bit of it [to something more fluid yet coordinated]. [The commanders] knew in the back of their heads what was going on in the logistics systems and the division communication systems, et cetera. So, the plan isn't a paper document. It's an intensive effort to understand all the moving parts and where you're trying to move the thing.

On an elementary level, to lead someone means guiding them—physically, psychologically, or cognitively—from where they are to someplace else. To do that, we have to know two things: where we are going (the vision) and what we must do to get there (the plan). Knowing what you are trying to accomplish and how you are going to do so is essentially the definition of having a plan. General Ed Rice said, "the number one job of a leader is to determine what should be done in that organization. [...] What's the future of this organization? Define that, and what are the things that we need to do to take us into that future?" Identifying

★★ ☆ ☆

the goal for those we lead and then creating a plan for accomplishing the goal are the primary responsibilities for any leader. Hopefully, through that plan or the ability to adapt it to the changing landscape, the leader and the organization can also accomplish the goal.

The plan is nothing; planning is everything

Though having a plan for accomplishing the mission is essential to leadership, because of changes in the environment, resources, or information, plans almost always change. German Field Marshal Helmuth von Moltke said, "No plan of operations extends with certainty beyond the first encounter with the enemy's main strength." Mike Tyson more concisely said, "Everybody has plans until they get hit." Put even more simply, plans change. Given that the environments in which we must lead are constantly changing, one could conclude that plans are useless. However, there is much more to having a plan, and that is the planning process. In General Frank McKenzie's experience, it is the planning *process* that is most instrumental to leadership.

> Having a plan gives you an intellectual framework that takes you forward in time. Planning is what you do when you've got a lot of time—deliberate planning. Time's not a factor. You can go back and argue a point. You can contest a hypothesis. You can have a dialectic. In combat, you don't have time for that. But if you consider that point before, you've got rich analysis that you can bring forward when you have very, very, very little time to make a decision. [...] That principle has guided everything I've done. Planning is critical. Eisenhower said, "The plan is nothing. Planning is everything." [...] I wouldn't go that far, actually. The plan is important, but planning is more important than anything else because that's [...] how you organize the future. [...] People who build good, solid, flexible plans tend to have options when you get into a time constraint. That's why planning is so important.

When we have invested the time in planning and have developed a clear plan for how to accomplish the goal, we have to know it well enough to be able to communicate it clearly to those we lead. In fact, they need to understand it so well that when unexpected-yet-inevitable changes (e.g., environmental, resource, information, etc.) occur, they can adapt within the plan and keep moving forward to accomplish

★★

the mission. Knowing the plan extremely well gives our people the freedom to improvise, as needed, without negatively impacting others. This is akin to great improvisational musicians. Such musicians are highly skilled, and they know the musical piece in exquisite detail. Their intimate knowledge of the music—the plan—allows them to deviate from it at times without disrupting the overall musical piece, usually creating something better than what would've been if the music were followed precisely. Similarly, when those we lead understand the plan completely, they can react rapidly to changing circumstances in ways that improve the group or organization's ability to accomplish the mission. General Skip Sharp explained how having a plan people understand, and can function within, leads to organizational success.

> If you as a leader understand what the mission of the organization is, where it needs to be going, how it needs to improve to be able to accomplish that mission, and the mission is a good mission and well laid out, and you have [...] the whole organization understanding what that mission is, and then [you are] empowering folks to do what they need to do to be able to accomplish that mission [...] then that organization will be successful.

WHAT IS A LEADER'S FIRST PRIORITY?

While General Glenn Walters was the Commander of the 2nd Marine Aircraft Wing in Afghanistan, a tragic accident occurred that left a squadron of his Marines demoralized. A V-22 Osprey squadron was just arriving in Afghanistan as the squadron they were to replace was preparing to return home. To help the arriving squadron's pilots become familiar with the area, the departing squadron would overlap with the new squadron for two weeks and fly missions with them. During one of those missions, General Walters explained that an outgoing master sergeant was asked to fly one more mission as the senior crew chief.

> The senior crew chief was a master sergeant who [because of having previously been on multiple deployments] did not have to go on that deployment, but his squadron commander said, "Would you extend (voluntarily prolong the length of service in Afghanistan) one more time,

because I really could use your leadership because of the young people going over?" So, the master sergeant said, "Yeah." So, he was a senior crew chief, and he ran the flight line for [the squadron commander, who was one of the mission pilots.]

The master sergeant was in the back of the Osprey, and when they reached their destination, General Walters recalled that he was unloading the cargo, comprised mostly of large, heavy-gauge plastic Pelican™ cases.

> So, he's doing that, and they had some [enemy] contact about a mile north of there, and a Huey and a Cobra (U.S. helicopters) were out there. [The Osprey pilots were] listening to the radio [and] said, "Well, we've got to get ready to get out of here, because they're probably going to launch some stuff here." So, they rolled up the throttles [...] the master [sergeant] was outside of the aircraft and unconnected [to the radio], but he heard [the engines rev]. He said, "Uh-oh." So, he jumps back on, and he can feel the aircraft starting to lift, [...] and they take off.

As the Osprey took off, the pilot surged the rotor power and pointed the nose up to make a quick departure. As he did, one of the heavy Pelican™ cases that had been unharnessed but not unloaded careened back through the cargo hold and knocked the master sergeant out of the aircraft from a height of about 120 feet, killing him instantly when he hit the ground in the landing zone. General Walters reflected on the impact of this tragedy.

> Because everybody knew him and liked him, [it] put the morale of that squadron down in the toilet, because it seemed so unnecessary. [...] it was the outgoing squadron who lost this young master sergeant. So, I had to go over and talk to them, and I talked about mission, and I told them what a tragedy this was. But, I had to take it back to the core mission that we're over there for. I said, "Remember, what are we doing?"

The squadron was heavily engaged in the assault on Sangin, which was a heavily contested combat zone on the way to the ultimate goal of Kajaki. If the squadron could not recover from the tragedy and resume flying, their fellow Marines on the ground would be endangered, and it would threaten the overall mission. General Walters grieved with the Marines in the squadron, and then he refocused them on the mission.

★ ★

I had to tell them [...] "I need you to get back online and start flying missions, because the young Marines out there, who've been out there for two weeks at a time, haven't had a shower. They need what you're going to bring them. They need you to transport them around the battlefield, so they can be successful."

The Marines in the squadron were able to reorient from their grief over the loss of their fellow Marine to focus on the needs of those Marines on the ground in combat, which allowed them to perform their duties and accomplish their mission.

Leadership only exists in circumstances where there is some goal to be achieved. That goal could be building a pyramid, improving the health of a group of people, selling more widgets, making it safely through a corn maze, or any other innumerable human endeavors. If there is no moving from point A to point B, if there is no need for change, if there is no mission, there is no need for leadership. This was a common theme amongst the Four-Stars, one emphasized by Admiral Mike Rogers who said, "We lead to achieve an outcome." If the primary reason for leaders to exist is to accomplish a given outcome/goal/mission, it follows that competent leaders are those whose groups or organizations accomplish it.

If the ultimate metric of leadership is accomplishing the mission, how can we use a structured approach to increase the potential for achieving it? Though none of the Four-Stars articulated a systematic approach to mission accomplishment, several discussed different steps leaders must take to be able to accomplish the mission. By analyzing their insights, I was able to synthesize those components into a systematic way of thinking about accomplishing the mission, which is presented in Figure 5.4.

The first step in accomplishing any mission is to **Know Your Mission.** That seems elementary, but untold numbers of endeavors fail because people don't really know what they are trying to accomplish. As a leader, you must know what it is, exactly, that you are trying to do. If you and those you lead do not have a clear goal you are trying to reach, you will not achieve it. Then, once you know what your goal is, to accomplish it you must **Plan For It**. Determine the barriers

★★☆☆

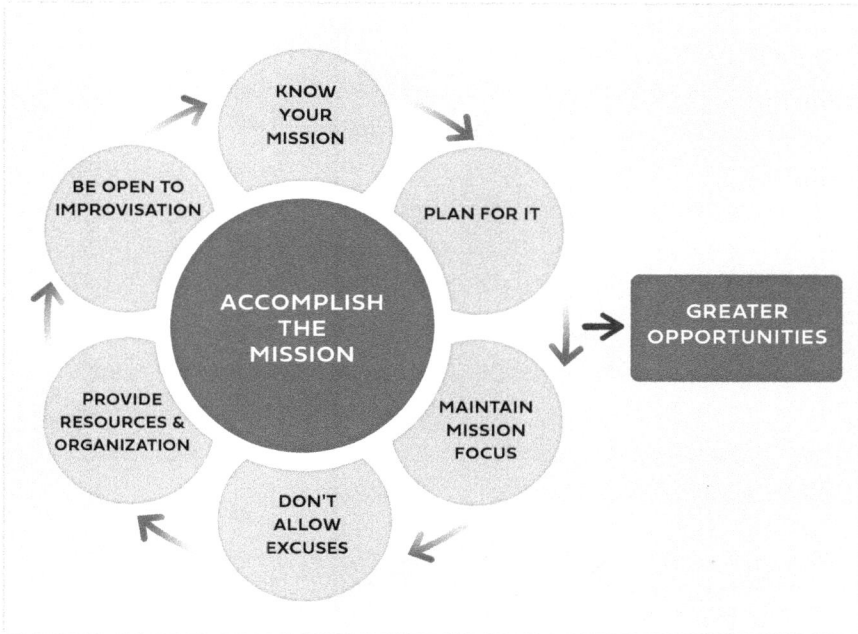

Figure 5.4. Systematic Approach to Accomplishing the Mission. To accomplish any given mission, you must know what the mission actually is, plan for it, maintain your focus on it, disallow excuses for why it cannot be accomplished, provide the necessary resources & organization so it can be accomplished, and be open to improvisation on the plan. When a leader and team follow this approach, the likelihood of mission success is greatly increased, and both are given greater opportunities.

to overcome (e.g., opponents, physical obstacles, financial and time constraints, etc.), as well as the resources (e.g., personnel, materials, time, etc.) needed to overcome those barriers and accomplish the mission. The next step in mission accomplishment is to **Maintain Mission Focus**; many things can distract a leader and the team, but to be successful, the mission must remain at the center of attention. General Paul Kern endorsed this, saying excellent leadership requires keeping a "mission focus [and] get[ting] the job done."

Excuses for why something cannot be done can erode mission focus. General Ed Rice admonished that leaders must not let that happen. He said, "leaders [...] have to be the person in an organization that doesn't allow the organization to find excuses for not getting something done."

★★

Effective leaders **Don't Allow Excuses**. Perhaps two of the biggest excuses for why something cannot be done are the lack of necessary resources and the organization to accomplish it. General Steve Lyons said a leader has "to have the ability to accomplish your mission and organize people and resources to do that." You must **Provide the Resources and Organization**; without those, few missions will be accomplished. Finally, General Stanley McChrystal provided a great insight into being successful when he said, "If it's stupid and it works, it isn't stupid. [...] The right answer is what accomplishes [...] your mission." Another way to frame this is a leader must **Be Open to Improvisation**, because successful plans are the ones that change to accommodate current and future circumstances. Successful leaders are the ones who harness the creative abilities of their people to improvise.

SELF REFLECTION

Once you have committed to a plan, how easy is it for you to change it? Are you more committed to completing your plan or accomplishing the mission? How do your prior actions support that?

HOW DO GREAT LEADERS RESPOND TO OBSTACLES?

In the early 1970s, then-Commander James Hogg was given an early command of the destroyer *USS England*, homeported in San Diego, CA. As his sailors were training for their upcoming deployment to sea, their engineering exam loomed—a series of highly comprehensive, in-depth inspections and interviews that stress-test the ship's engineering systems and determine how well the crew is trained to go to sea. If the crew did not pass the engineering exam, they would not be allowed to set sail. Commander Hogg was reminded of that daily as

★★☆☆

he saw the destroyer *USS Fox* anchored alongside the *England*—they had failed the exam and weren't going to deploy. Commander Hogg thought to himself, "That cannot be us. We cannot be tied up alongside a peer designated unable to steam because we didn't pass our engineering exam."

Two weeks before their scheduled deployment, Admiral Hogg shared, the *England* had their exam.

> During the exam, everything went pretty well. We were well prepared for it. The officer who conducted the exam said, "Yes, you passed." He sent a message out to the fleet commander saying we passed. I gave the engineers a case of champagne, and off we went [to celebrate].

The crew was energized because they had passed the engineering exam, Admiral Hogg recalled, and they celebrated into the night.

> The next morning, I got to the ship, and I had a call from this inspector. He was a commander, and his boss was a captain. [...] He said, "Captain so-and-so has decided that you didn't really pass because there were a couple of things that weren't quite right that have to be corrected." One of them was a stress test of the boiler, [and] another one was [the engine's] valves...

The message went out to the crew that they had failed the inspection, and they were dejected because it meant they wouldn't be deploying to the Western Pacific as planned. However, Commander Hogg was determined to find a way for the *England* to deploy on-time.

> I had two weeks before deploying. I got right on the phone. I said to him, "Look, tell me what things we need to do in order to pass a re-exam before we deploy." He said, "Well, I don't know that you'll get a re-exam." I said, "Well, I'll take care of that with my [Navy weapons system] type commander. You just tell me, 'If you do these things...' For example, if our valves are not in good enough condition, I can't repack every valve in the engineering plant, but [...] I can repack about 10 valves, and I can establish a valve repacking system with all the equipment in it to demonstrate that we can repack all the valves over time properly." He said, "Okay, that will do. Oh, also, there was a little bit of dirty stuff behind one of the boilers." I said, "We'll take care of that."

★★

Commander Hogg had gotten a window of opportunity for the *England* and her crew. "I knew we could do it. I believed in it, and I got the crew to believe in it." He called his type commander and squadron commander and got their support for the plan.

> I went over to the waterfront, to the engineering readiness group that helps repair ships, to work on the boiler stress test. A week later we got it away, went to sea, passed the test. As far as the cleanliness and sanitation of the engineering spaces, the whole ship volunteered to go down and help clean it up. That happened in one day, [there were] that many guys working down there. We passed, and we sailed to WESTPAC (the Western Pacific) on time. In WESTPAC, we were deployed for eight months. The fleet commander said we had the best operating tempo of any ship that had ever sailed to WESTPAC. [We] never had casualties, met every commitment, [and] got a lot of kudos. [We had] very high morale.

Few things worth accomplishing are easy. If they were, they most likely would've already been done and wouldn't require leadership. That is to say, challenges come with the territory. Overcoming those challenges may be the entire mission, or the challenges may arise along the way as obstacles to accomplishing the goal. Four-Star leaders recognize and embrace this reality. Instead of allowing themselves or those they lead to make excuses for why a difficult mission can't be accomplished, the best leaders maintain a "can do" attitude. That attitude was at the heart of one of General Ed Rice's leadership maxims: "Find a way or make a way."

> You're going to have obstacles. The more senior you are as a leader, the tougher the challenges or the obstacles. Difficulty is a challenge; it's not an excuse. You've got to find a way or make a way.

General Rice was one of many Four-Stars who discussed and gave examples of the power of the "find a way or make a way" mindset. Another was General Ed Eberhart.

> [If you are a leader, you can't] stand around and complain about the cards that were dealt to you. You figure out how to best play those cards. [...] You've got to be solution-oriented. You're confronted with problems day-in and day-out that you must solve. It does no good whatsoever to sit around and [complain] about it.

★★ ☆ ☆

Four-Star leaders know that when they and their group maintain a "find a way or make a way" orientation, seemingly insurmountable obstacles can be overcome. "Find a way or make a way" carries an implicit permission toward improvisation, innovation, and creativity when necessary. When talented and well-trained people are given the freedom to improvise in the face of an extreme challenge, they can almost always find a way or make a way.

A major way that Four-Star leaders are able to "find a way or make a way" is by having a bias for action. That is, they have a personal predisposition to take action—to make the future they want instead of waiting and hoping it will simply unfold. They take enough time to assess the situation and make a good decision based on the information available to them, and then they take action. Nearly three-quarters of Four-Stars either specifically discussed or demonstrated through examples how having a bias for action helped them accomplish their goals and opened up further opportunities for them.

General Keith Alexander, *on Having a Bias for Action*

In 2003, then-Lieutenant General Keith Alexander was named the Deputy Chief of Staff for Intelligence for the U.S. Army. Thereafter in 2004, the story came to light of the heinous acts perpetrated against prisoners at Abu Ghraib. The management of Abu Ghraib had been under the auspices of the Army's Deputy Chief of Staff for Operations (G3); however, when the news broke, Lieutenant General Alexander found things abruptly changed, and he had to act fast.

> The G3 said, "Interrogation operations are an intel function, over to you." I got passed that in the morning, and I was told there was a meeting with [Secretary of Defense Donald] Rumsfeld. So, I quickly got the interrogation manual out [and] read through it again real quick. I walked up to the meeting and said, "Hey, I'm here for the meeting." [The G3] goes, "Are you ready?" I said, "Ready for what?" He said, "You're the briefer. You're briefing Rumsfeld." It was a one-hour briefing on interrogation operations. I thought, "Well, I've got two more minutes to prepare." So, I had just read through it, and I walked him through the briefing, how you do interrogations. [...] He said, "General, I need you to go to Congress and brief them in the hearings that are coming up Thursday."

★★☆☆

Lieutenant General Alexander's bias for action put him in a position to know his stuff when it came time to brief the Secretary of Defense. He didn't sit idly by, but rather he took action to learn the interrogation operations manual, which prepared him to deliver during the briefing. His impressive performance led to the opportunity to brief Congress. As will be covered in more depth in Chapter 12, Lieutenant General Alexander's knowledge of the situation at Abu Ghraib and familiarity with the members of Congress allowed him to explain the situation effectively to them in a way they could understand it. His performance before Congress put him in good stead with the Secretary of Defense and numerous powerful members of Congress, which set the stage for his promotion to Four-Star General and being named the Director of the National Security Agency.

★★☆☆

CHAPTER 6

TODAY'S LEADERSHIP IS NOT GOOD ENOUGH FOR TOMORROW

You are a builder, a developer, a grower of capacity of your organization. [...] You're improving yourself all the time. You're growing your capacities and everything. What you were yesterday is not good enough for tomorrow. It's one of constant growth of yourself. You're also developing and growing your leaders all the time. It's a constant growth because things are changing fast...

– GENERAL DAVID RODRIGUEZ, U.S. ARMY –

As a Three-Star General with over 34 years of leadership training and experience, then-Lieutenant General Vince Brooks had reached rarified air with regard to his leadership capabilities. Yet, he was not finished; he knew there was always more he could learn that would help him improve as a leader. As part of his ongoing military leadership development plan, General Brooks participated in "Leadership at the Peak," a program offered through the Center for Creative Leadership.

It's a very, very intensively focused, week-long program for business executives principally, and a few military leaders who were already senior in their field. [...] They were already CEOs, or COOs, and CFOs, Three-Star Generals. [...] It was one of the most impactful courses that I ever went to.

★ ★

Up to that point, General Brooks had been highly successful as a military officer, decorated for his leadership and for commanding over 100,000 soldiers in battle. Though his leadership had been forged in the harshest of battle environments, his learning experience at Leadership at the Peak was rawer and more emotional than might be expected for a hardened soldier.

> What they helped me to understand at that course is to get to the next level of performance, when you've already accomplished and reached the highest levels of attainment, [...] to reach the highest levels of performance, you have to look within. Everything from: How are you fueling your body? How do you recharge it? Where are you in your own way in terms of leadership that has worked for you in the past, but is less effective than you think it is?
>
> [Professional coaches] put us through a variety of scenarios that week, [and] did some very deep, insightful things to get to the deepest level of you. By the end of the week, I'm not kidding you here, most of us were crying in the last exercise, which was very self-introspective. Writing a letter to yourself that you would review 60 days later, the emotional outflow that came from that was surprising, and you don't care at that point. You've been with people for a week, you've been deliberately placed into vulnerable positions so that you can get past yourself a little bit, and then you can get into the real level of learning.

General Brooks reflected on the program's impact on his leadership, as well as his life, and why being a great leader requires ongoing, career-long training and leadership development.

> As a result of that [course], I was a more effective Three-Star and Four-Star General than I would've been had I not gone to that, without a doubt. I'm a dietary vegan now, for example. That came out of that course. It's just one small thing, it's a lifestyle change, but it's all about how I fuel my body and my brain. [Being a successful leader requires] continuously working to improve yourself, never being satisfied that your leadership is done, [or thinking] you've got it exactly right; you don't ever have to get any better.
>
> I often thought, "Wouldn't I have been much better off if I had taken this earlier?" And I probably would've. But I think the context within which I was learning about myself was very different because it was when I was a mature leader. When you're still trying to figure out your leadership

★★☆☆

style, and you're just getting your first few experiences of success or failure under your belt, you're not sure that you're crystallized in what your leadership is. But when you're very senior, you might be crystallized in what you think is your leadership, only to be shocked into realizing that you're not finished yet.

Any leader at any level must be aware that things are constantly changing. New challenges and threats arise continuously. The ever-changing landscape demands that we must be constantly changing, evolving, and growing, as well. No leader has the skillset to lead successfully in every situation, but most can learn how to do so in any situation. Those leaders who cannot or refuse to keep learning and growing in their capabilities are doomed for mediocrity at best and utter failure otherwise. They will watch their leadership capabilities fade and themselves become irrelevant.

As quoted at the opening of this chapter, General Dave Rodriguez framed the idea succinctly: "What you were yesterday is not good enough for tomorrow." Four-Star leaders know they must be continually improving, constantly growing their capabilities, because what got them to where they are isn't going to get them to where they want to be. The challenges that were overcome to get to today will be harder tomorrow. General Chuck Wald said, "You've got to continue to grow every day. [You've] got to keep working at it. If you're on top [...], you better keep swinging because they're coming after you." Your competition will be smarter, faster, and stronger tomorrow. You must be as well if you are to be successful. As is often said amongst Navy SEALs, "The only easy day was yesterday." If you want to be a Four-Star leader, embrace the reality that leadership is a continual growth process, a mindset of being dedicated to learning and improving. General Jim Mattis continues to be better tomorrow by seeing today as an opportunity:

> Every day is a school day. You're learning every day as a leader, so you're never too old not to make mistakes. You're never too young to have a good idea. And by the way, just make sure that you do a lot of good listening in order to have a school day. We listened to our teachers, and my teachers, even in my last days, went from age 18 to age 97. Those were my teachers.

★★

Throughout his career, even as a Four-Star, General Mattis maintained the attitude that he should continue to learn every day—a growth mindset. Importantly, he did so by recognizing that everyone (i.e., "age 18 to age 97") has something they can teach us if we are willing to allow ourselves to learn from them. If we are going to be the leaders that tomorrow demands, we must continue to grow in our capabilities today.

WHAT DOES IT TAKE TO GET BETTER?

As General Bob Magnus progressed through a 37-year military career that culminated in serving as the Assistant Commandant of the U.S. Marine Corps, he realized the important role failure plays in learning.

> Failure is normal. Everybody should realize there are consequences to their decisions and actions, and obviously some failures are less forgivable than others. [...] You have to let people stumble as they learn to walk. [...] You can't excuse it, but [you] use [it as] a teaching moment. [As an analogy, it is] the idea that if you're going to be an Olympic gold medalist, [realizing] not clearing the horse on a leap is [not] fatal to your position on the team [...] [and having the mindset] of, "Let's see what went wrong here, and how to do it right."

Thinking back on his career, General Magnus recalled the story of an officer who made a significant mistake as a leader in combat, but because his commander saw it as a growth opportunity, that officer learned something important and went on to a storied career.

> I know a very famous Four-Star General, who in combat at a battalion level was very, very good. [...] As they were moving quickly past an enemy position that had been eliminated, his formation of troops behind him was ambushed by enemy troops that were in the position that had supposedly been taken and suppressed. Of course, this is the chaos of battle... So, bottom line, he allowed his unit to be surprised and ambushed. That wasn't fatal; I mean, it might have been for one or two individuals [and] certainly the enemy.
>
> His regimental commander came up to him immediately after this took place, within single-digit hours, and put his arm around him and said, "Did you learn something today?" This was a good unit commander, and he actually did. He learned that maybe you can bypass, but then

★★☆☆

you have to protect against that. [This demonstrates the value of] giv[ing] people [...] the latitude to fail within bounds, and when they fail within reason, you use that as a way to help them grow and help [...] their peers grow. You don't hold them up to ridicule or something like that. When they break the bounds, then you have to be reasonable about what you do, because you can't have the lesson that failure is not tolerated...

A striking characteristic among the Four-Stars is a dedication to continuous improvement, increasing competence, and seizing failures as learning opportunities. Instead of believing their skills and abilities are set in stone and any failure, shortcoming, or mistake on their part reflects inherent, uncorrectable defects of character or ability— an outlook known as a fixed mindset—excellent leaders are typified by seeing mistakes and shortcomings for what they are—evidence of something they have not yet mastered, but something they can learn to do. This mental outlook is referred to as a growth mindset, and it is an invaluable leadership resource.

Carol Dweck, PhD is a renowned psychologist who has identified, studied, and explained the growth mindset. Perhaps one of the easiest ways to understand the difference between the fixed and growth mindsets is to think about the age-old leadership question: Are leaders born or are they made? Someone with a fixed mindset would likely believe leaders are born—either a person has the ability to be a leader or doesn't. For a person with the fixed mindset, they may progress along a leadership arc being successful until, for whatever reason, at some point they run aground and have a leadership failure. Instead of seeing that situation as one in which they still have something to learn, their leadership failure translates to them being a failure—not having what it takes to be a leader—because in that mindset, either you have it, or you don't. Additionally, people with fixed mindsets often avoid challenging situations because of the risk of failure; if they were to fail, it would force them to see themselves as failures—if they never try anything difficult, they are unlikely to fail, which allows them to protect their self-image. Conversely, a person with a growth mindset knows that leaders are made, and they are made over time through learning from their shortcomings and mistakes. When a leader with a

★★

growth mindset has a leadership failure, they see it as an opportunity to learn. Unlike those with a fixed mindset, a leader with a growth mindset is more likely to take on challenging situations because they view these situations as opportunities to learn, not threats to their self-identity. Figure 6.1 presents some of the contrasts between the fixed and growth mindsets.

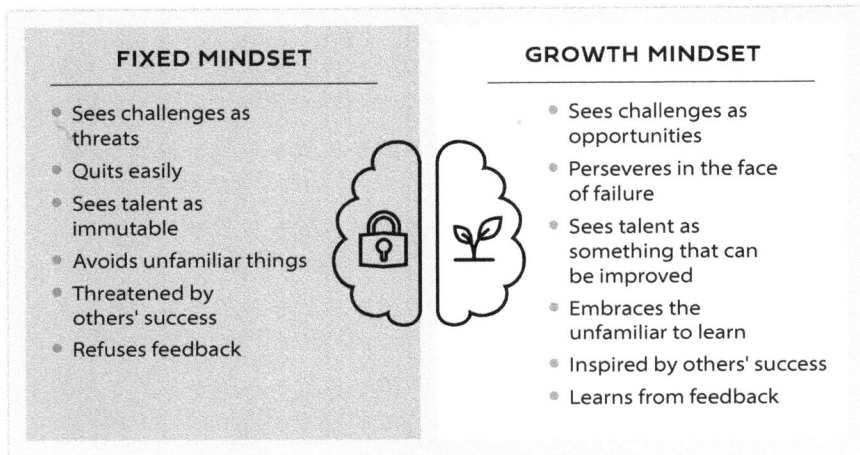

FIXED MINDSET	GROWTH MINDSET
• Sees challenges as threats	• Sees challenges as opportunities
• Quits easily	• Perseveres in the face of failure
• Sees talent as immutable	• Sees talent as something that can be improved
• Avoids unfamiliar things	• Embraces the unfamiliar to learn
• Threatened by others' success	• Inspired by others' success
• Refuses feedback	• Learns from feedback

Figure 6.1. Comparison of Fixed and Growth Mindsets. A person with a fixed mindset avoids challenging situations and unfamiliar things because they believe their abilities are locked in place. Failure at an endeavor is equated to being a failure as a person. Conversely, a person with a growth mindset seeks out challenging and unfamiliar things because they see them as opportunities for growth.

Learn from failure

We expect leaders to know things, to be able to do things, and to be successful. Mistakes, shortcomings, and failures aren't supposed to be part of the leadership story, or so it is usually believed. Our tolerance for leaders' mistakes is low because they threaten the mission and the group or organization as a whole. Certainly, there are failures by leaders of such magnitude that they require the leader's removal. However, rarely do such failures happen without numerous prior mistakes and failures from which those leaders did not learn and which led to continued errors, often of increasing severity. Additionally, because of the tendency to demand infallibility, when leaders fail, in an effort to avoid

★★☆☆

negative repercussions, those leaders may either ignore, hide, or explain them away. As a result, they don't learn from them and are, in some ways, a less capable leader than they were before the mistake.

Instead of ignoring, hiding, or explaining away mistakes, Four-Star leaders see failure for what it is—an opportunity to learn. General Steve Lyons sees leadership as "a learning environment." He said, "The leadership journey is all about adapting and learning and being resilient," and being resilient means there was something that had to be overcome—failure. It is cliché to say we always learn more from failures than from successes. In fact, it would be more accurate to say we seldom, if ever, truly learn from successes. When we are successful, we get something right; we already know what to do. Where is the learning in that? We only learn when there is something we don't know or can't do. To learn, grow, and improve, we must learn from our failures.

To understand that failure is the gateway to learning requires that we shift our mindset, both how we view ourselves and those we lead. To learn from our failures requires that we embrace them as an opportunity to improve—we have to own them. Admiral Scott Swift said, "Failure is our best teacher as long as we own those failures ourselves, and we don't blame those failures on subordinates, or we don't blame them on the organization." By owning our mistakes and having a growth mindset that allows us to recognize them as opportunities to learn and improve, we can grow in our ability to lead. The U.S. Marine Corps demonstrated this growth mindset when they continued to invest in a young Marine officer named Jim Mattis, who acknowledged the effect it had on him.

> I made tactical mistakes. Some of my mistakes cost troops their lives, and certainly could have cost a lot more, other than, thank God, for the NCOs and junior officers who got me out of the mess I'd gotten them into. But I think it was to the Marine Corps' credit that they realized I was learning what right looks like by making mistakes.

We all make mistakes, and the best leaders seize those mistakes as learning opportunities in becoming better leaders. Additionally, Four-Star leaders recognize that mistakes can come at a great price, so those opportunities to learn from them must never be wasted.

★★

HOW CAN LEADERS BE BETTER TOMORROW?

Including his four years as a cadet at the United States Military Academy at West Point, General Scott Wallace spent 43 years in uniform, and throughout, ongoing learning and improvement was a part of the *ethos*.

> It's ingrained in the organization. So, it's second nature. We have within the Army, a huge infrastructure that's built to [train people continuously]. We have combat training centers [...] We've got a simulation-based training regimen for division, corps, and senior-level commanders and their staff through the Mission Command Training Program. Those events create a culture that not only expects to train, but they are also very open and frank about what they did right and what they could have done better as an organization.

General Wallace's commitment to being better tomorrow not only provided him the capabilities to rise to Four-Star rank, but it also gave him the opportunity to command the United States Army Training and Doctrine Command (TRADOC), which oversees the training of all the Army's personnel. Throughout his career and during his time leading TRADOC, he saw and employed the after-action review process to help leaders continue learning.

> I spent six years at the National Training Center at Fort Irwin, and we were always very proud of the after-action reviews that we conducted with the commander and staff and subordinate commanders. The gist of the discussion was, "What happened, why it happened, what could we do better?" We had instrumentation and overhead pictures, and all kinds of stuff that we could show the organization [so they could see] exactly what happened digitally and sometimes on film. So, when you get past what happened, all the arguments go away, because you can show what happened. Alpha Company screwed up here. Here's how that impacted the overall battalion's operation. Okay, let's set that aside. Why'd that happen? What decisions did the commander make that caused that to happen? If given the same situation some other time, what decisions would you make in lieu of the ones that you made?

When he considered the question of how to get others to see the value of continual learning and focusing on being better tomorrow than you

★★☆☆

are today, General Wallace said, "It's hard for me to answer your question, because it's never been a question. It's what we do."

As General Wallace pointed out, ongoing professional development is woven into the fabric of the military. That's because the world is constantly changing. The demands of leaders are constantly changing. People change, societies change, processes change, resources change, and environments change. With ever-increasing advances in communication, computation, and logistical coordination, the velocity of those changes has become exponential. Four-Star leaders seek to be continually learning, to be better tomorrow. They do that through personal study, formal training, designed on-the-job experiences, and seeking the counsel of mentors who've been there. These four development processes comprise what has been found to be the most effective leadership development systems.

SELF REFLECTION

What are you doing today to be a better leader tomorrow?

General Brooks, *on Always Improving as a Leader*

General Vince Brooks believes strongly that leaders should always be improving, growing, and advancing in their capabilities.

> Never feel satisfied with your own leadership. You can, in fact, be a better leader. No matter how long you've been doing it, or how large an organization you've been placed responsible for, you can still get better. [...] [Ask yourself,] "Can I be a better leader tomorrow than I was today? What did I learn about my leadership today? What worked, what didn't work?" [...] You've got to be on your game. You've got to keep sharpening it. You can't just do things the same old way, because your people need something, or the circumstances have changed.

★★

WHEN DO GREAT LEADERS STOP LEARNING?

General Frank Grass served as the 27th Chief of the National Guard Bureau, a position in which he commanded nearly a half-million military members and oversaw a budget of $25 billion dollars. Long before that role, however, he realized if he was to improve his leadership ability and opportunities, he had to dedicate himself to continual learning. When he progressed through the leadership ranks to become a colonel, he foresaw his risk of falling victim to the Peter principle. He recognized that as an engineer there were many things that he didn't know about leading other groups that would be crucial if he hoped to move on to become a One-Star General and beyond.

> I've got to [...] get out of the mode of being that engineer. I'm now becoming a generalist. I've got to look across the force. I've got to understand all the capabilities. I've got to understand the Joint world. For me, understanding the Air National Guard and the Army Guard became very critical. But, [during] my time at Northern Command, I had to understand the Coalition Force because we had Canadians there. We had the Army, Navy, Air Force, Marines, Reserves, Guard, civilians. All the agencies with the United States government were represented there. I had to just open my mind up and be much broader and do some studying. How lifelong learning I think really helped me a lot, especially working in the combatant commands [...] is studying these organizations, their culture, how they function, what kind of resources they have, different organizations and even to a point where I used to pick executive officers to work for me [...] from a different service most of the time because that was part of my education about understanding culture.

General Grass made a concerted effort to learn as much as he could from members of other services, because he recognized that in a high-level leadership role, his scope of understanding needed to be much broader.

> [I realized I needed] to better understand the big picture of the U.S. military if I'm going to be a general, because there's a good chance I'm going to end up in a lot of different positions. [...] When you're going through the ranks at the lower grades, you try to be very proficient in the tactical side of your mission, in the operational side. But, now that you're at the

★★☆☆

pinnacle side of being a senior leader as a general officer, and especially into the Three- and Four-Star, you have to spend time thinking and reading. You've got to go off and think about what does leadership mean? How do I lead? What are my strengths and weaknesses? How did I shape my staff based on my strengths and weaknesses? What is the culture of an organization like the National Guard Bureau? [...] So again, lifelong learning is critical for a senior leader.

In addition to the role lifelong learning has on being able to lead in the current environment, General Grass also emphasized its value in being able to respond to the ever-changing environments in which leadership must occur.

> The other thing I think today more than ever, especially in my career and especially in the last 20 to 30 years, is the whole world changes so rapidly, and you have to stay up. You have to keep yourself educated in the changes. It gets hard as you get older because you're getting set in your ways.

In constantly changing environments, the only way to remain effective as a leader is to be better today than yesterday, which necessitates ongoing learning. Four-Star leaders recognize and embrace continual, lifelong learning as a requirement. When the world is continually changing, so must the leader by learning and growing. This was a common theme amongst the Four-Stars. For instance, General Steve Lyons said, "Leadership is perpetual learning." Similarly, General J.D. Thurman insisted, "Learn. Never stop learning, and encourage others to do the same," which complimented General Gene Renuart's conclusion that "Good leaders never stop learning." The Four-Stars made it clear that to be excellent leaders we must be learning constantly, because everything else is changing constantly. For them, it is a matter of competence. Developing your leadership ability is "the quest to keep learning," according to General Mike Scaparrotti. "You've got to continue to learn, or at some point you just don't know what you need to know to be competent." For those who think they have leadership figured out and refuse to keep learning, General Lori Robinson had a stark warning—"The nanosecond you think you know everything, you've failed." Four-Star leaders always keep learning.

★★

The return on investment in personal leadership development

What does lifelong learning look like? For military leaders, it means a formalized development plan with three major components: formal educational training programs occurring at regular intervals, periodic changes in leadership position and context, and receiving professional advice through mentors. These three components represent knowledge, experience, and mentoring/guidance and are the basis for any structured, personal leadership development plan. As represented conceptually in Figure 6.2, there is a direct relationship between how much effort we put into developing our leadership capabilities and the leadership competence we achieve—the more committed you are to developing your leadership capabilities, the better you will be.

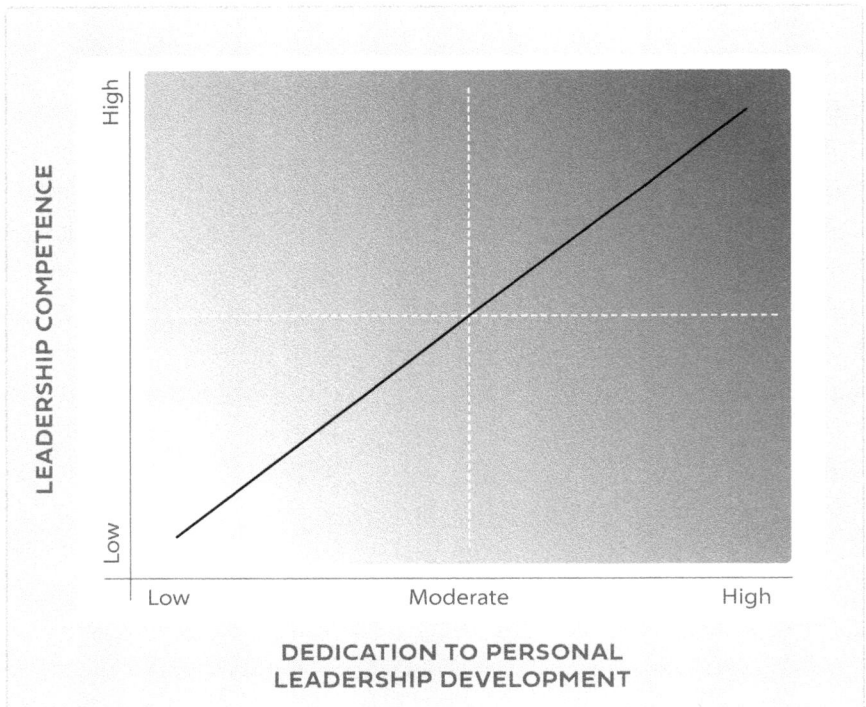

Figure 6.2. Relationship Between Leadership Competence and a Leader's Dedication to Personal Leadership Development. The more dedicated a leader is to investing personal effort in developing as a leader, the greater that leader's competence will be.

★★☆☆

While most people outside of the military do not have formalized career development plans, the principles from the military are translatable to any domain. Specifically, to continue to learn, anyone can work to craft their own personal leadership development path comprised of those three principal components (i.e., knowledge, experience, and mentoring). This may include formal training to acquire desirable skillsets, such as through graduate training programs (e.g., MBA, MS, DSL, etc.), coaching training, various certificate programs, or online courses. It could also include informal study through structured readings over any of the myriad books available. Additionally, leaders can seek out opportunities in their current roles for further responsibility and professional experiences to broaden their skillset. Finally, any leader who hopes to continue to improve should seek out and engage in ongoing mentoring from someone who has been there. This may seem like a big request of someone, but it is surprising how willing people are to tell you how they accomplished something if you are only willing to ask.

WHERE DO FOUR-STAR LEADERS TURN FOR ADVICE?

There is a great deal to be learned from both formal and informal development endeavors. However, those components that focus principally on gaining knowledge and theory lack the opportunity for practical application. Conversely, experiential activities, absent guidance or nuance, come with the risk of failures that, if substantial in magnitude, could lead to career setbacks. Recognizing this, Four-Star leaders take advantage of the wisdom and experience of mentors who have been there—who have led in the types of environments, contexts, and situations the leader is currently facing or is likely to face in the future. By doing so, Four-Star leaders can accelerate their own knowledge and capability by learning from the mistakes and lessons of their mentors. This was something General Keith Alexander concluded as he thought about the most important keys for leadership success, saying

★★

that "having a mentor or a set of mentors" is instrumental. It is the wise leader who learns from the successes and mistakes of others.

While multiple Four-Stars recommended the necessity of having "a mentor who has been there," good mentors and mentoring relationships can be difficult to find and cultivate. Numerous studies have reported a lack of effective and knowledgeable mentors across multiple professional domains. However, drawing from my personal experience of having had numerous exceptional mentors, having studied mentoring, having served as a mentor to more than 50 physicians, and having built and led mentoring programs, I have developed guiding principles for both how to identify a mentor and also how to make the most of the mentoring relationship.

I have identified six guidelines for finding a great mentor. The first guideline is to search for one. Great mentors are like gold—while they might occasionally be in plain sight, most of the time you have to dig around. A principal reason people fail to find the right mentor is because they don't invest the time and energy to do so. There are three questions that can serve as tools for unearthing a good mentor: Who has a lot of successful mentees? Whom do people seek out to be their mentor? Whom do people avoid?

The second guideline for finding a great mentor is to identify the skillset you are hoping to acquire. Once you know what you are wanting to develop, find someone who is an expert in that skillset and, particularly importantly, who can teach it. If you want to be more technically adept, then you will need a mentor with great technical expertise. If you want to be a better communicator, you will want someone who is skilled at communication. Notably, many leaders mistakenly think they must have a mentor who is in their exact same technical field. As it pertains to leadership, that is almost always incorrect. As this book lays out, Four-Star leadership is about Character, Competence, Caring, and Communication, none of which is specific to a given technical field. So, it is worthwhile to explore potential mentors outside of your specific professional domain.

There are a set of personal characteristics that have been identified in the ideal mentor, and finding someone with those is the third

★★☆☆

guideline. In a 1999 study of mentors across five organizations, Tammy Allen and Mark Poteet identified five key characteristics of an ideal mentor. They should have excellent listening and communication skills. Patience is a virtue among mentors. Great mentors have knowledge of the organization and industry. Ideal mentors should have emotional and social intelligence—the ability to read and understand others. Finally, a model mentor is characterized by honesty and trustworthiness. An absence of any of these characteristics can negatively affect the mentoring relationship.

The fourth guideline in finding a great mentor is to identify someone with a willingness to invest in your development. In an ideal mentoring relationship, the benefits flow in both directions: the mentee receives guidance and development, and the mentor receives fulfillment in helping someone else grow in their capabilities, as well as occasionally being able to foster their own career growth. Unfortunately, there are those would-be-mentors whose focus is only on building their careers. These are the people you want to avoid, and this is the benefit of asking around to see whom others recommend and avoid.

The next guideline in identifying the right mentor is fit—is this someone you feel like you can get along with and learn from? Sometimes it can take a while to determine if there is a good fit. One of my own mentors used a simple litmus test when it came to this idea. He always asked himself, "Is this someone I'd like to have dinner with?" If the answer was yes, then things usually worked out well. If the answer was no, he would pass. This simple question has helped me tremendously with regard to finding mentors, as well as knowing when to pass on mentoring someone.

The final guideline for identifying a great mentor is a derivative of the previous, and it is knowing when to walk away. There will be potential mentors whom you spend time getting to know, perhaps even working with for some period of time before it becomes clear that because of a lack of fit, an absence of the characteristics or expertise needed, or some other reason, the relationship isn't going to be productive. In such situations, the best course of action is to end the

★★

mentoring relationship as amicably as possible. Unless the person is Machiavellian, they will almost always be thankful you did.

Having addressed how to find a great mentor, the other component of having a successful mentoring relationship is how to capitalize on it once you have it. Doing so can be boiled down to two fundamental principles: respect your mentor's time and investment in you, and be dedicated to your own personal development. To expand on that, I've identified six valuable practices for getting the most out of mentoring, and these are presented in the table.

Table 6.1. How to Make the Most of Being Mentored

MENTEE ACTION	WHY IT'S IMPORTANT
1. Make time for it	To get anything out of mentoring, you must invest the time
2. Set clear goals and expectations	Both the mentee and mentor must know the goals and expectations; what are the deliverables & who's responsible?
3. Prepare for mentoring meetings	If you don't prepare, you're wasting everyone's time and ensuring the meetings will be less beneficial
4. Be consistent and committed	If you cannot commit to your own development, you'll never be able to lead others successfully
5. Seek to be challenged	Receiving and acting on challenging feedback is rocket fuel to your development
5. Be open to new ideas, approaches, and unconventional solutions	A great mentor has a wealth of experience that almost always exceeds yours. Capitalizing on that will greatly increase your success

★★ ☆ ☆

HOW DO GREAT LEADERS AND ORGANIZATIONS REMAIN COMPETITIVE?

In October of 2018, after 39 years of service in the U.S. Marine Corps, General Glenn Walters left his position as the Assistant Commandant to become the President of the Citadel. For the first 18 months in his new role, he faced many of the same challenges a lot of university presidents face, and then something unprecedented arose—the COVID-19 pandemic. With colleges and universities across the country shifting to online classes and being unable to retain significant numbers of employees, the Citadel was faced with the same challenges. At that point, General Walters chose to go a different direction, one informed by the Marine Corps maxim "Improvise, Adapt, Overcome." He recounted, "In this school, the one decision that I made that was probably good during the pandemic was, I decided not to furlough anybody, and I wasn't going to take an across the board cut. I mean, all the schools in this state did. What I did [instead] was repurpose [the staff]."

General Walters looked at the capabilities of his people across the Citadel campus and identified ways to both keep the people employed and to provide service to the community.

> I had my tailor shop instead of doing uniforms, because all cadets weren't here, they were making masks. Our innovation center, they started making N95 masks, and the printers, and we gave them to the hospitals. Our laundry did laundry for the places out in town that needed the laundry to be done. I'm talking about the shelters and homeless. [...] I rented the barracks out to the Marines for Parris Island.

In addition to the staff of the Citadel being able to continue working and generating income for the university, there were valuable second-order effects from General Walters' decision to adapt to the situation.

> We helped out the community that way. I think it kept our morale up a little bit, because they knew that everybody else was [being laid off]. In fact, we have some of our campus security folks, there's three of them here that got laid off by other schools. So, they came down here. So, we got that benefit, and I communicated that, and tried to make our own luck on campus.

★★

In the face of the continuous changes that render today's leadership inadequate for tomorrow's demands, adaptability is essential for leaders. However, it's valuable to differentiate adaptability from flexibility. Flexibility is the characteristic that allows something or someone to momentarily bend to accommodate a given force or situation. Adaptation, on the other hand, is the change or the process of change through which an organism becomes better suited to its environment. So, adaptability is being able to change what we do and/or who we are so we can thrive in a new setting.

Adaptation can be particularly challenging because it demands meaningful, fundamental change of the sort that usually challenges what we have been doing and the things we hold dear. If those are now inadequate or wrong, then, absent the most resilient growth mindset, our competence is called into question. This can threaten our self-image, which leads to a defensive posture and resistance to change. As a result, it can be easy to slip into rigidity, which undermines our ability to improve, as well as to function well as a leader.

Those who can overcome these cognitive roadblocks and are able to be adaptable greatly enhance their potential for leadership success. General Mike Scaparrotti credited his own adaptability for being the reason he advanced to the level of a Four-Star General, saying, "I wouldn't have made it even to Two-Star or beyond had I not been willing to change some of my habits and learn." He went on to say, "If [leaders are] not willing to work and change, they're not going to reach [their] potential and they can't serve [...] in the way that we would like them to serve." Admiral Pat Walsh said that to be successful, a leader must "find a way to adapt to a new set of circumstances and to understand what it means to be effective." In fact, nearly 75% of the Four-Stars identified or discussed adaptability as being instrumental to successful leadership.

Beyond recognizing the importance of adaptability to leadership, what are the characteristics required for one to be adaptable? I identified at least four contributors to adaptability based on the Four-Stars' interviews. (Figure 6.3) To be adaptable, a leader must first possess what General Jim Mattis referred to as "a willingness to be persuaded."

★★☆☆

That is, as a leader, you need to be able to hold in mind that you could be wrong or that you may not have the best plan. This can be particularly difficult to do, especially if you approach a given challenge with the mindset that you must be the one to come up with the solution.

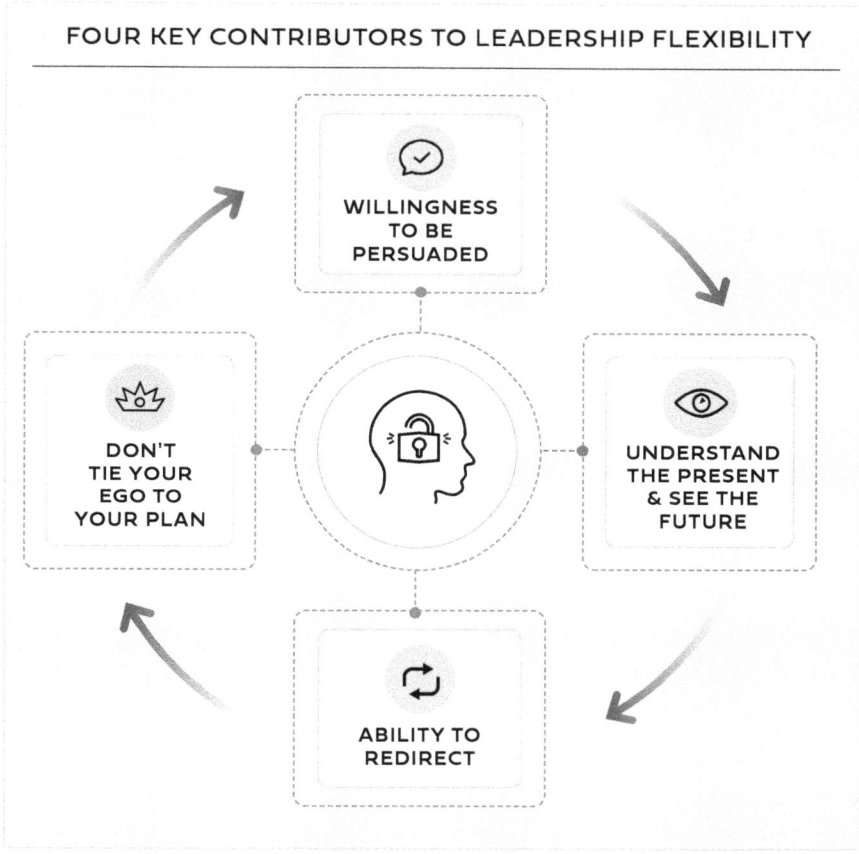

FOUR KEY CONTRIBUTORS TO LEADERSHIP FLEXIBILITY

WILLINGNESS TO BE PERSUADED

DON'T TIE YOUR EGO TO YOUR PLAN

UNDERSTAND THE PRESENT & SEE THE FUTURE

ABILITY TO REDIRECT

Figure 6.3. Four Key Contributors to Leadership Flexibility. Four-Star leaders maintain a willingness to be persuaded that someone else may have a better plan. They are typified by the ability to understand their current environment, the needs to perform in it, and where things are moving. They have the ability to redirect from their plans to something more likely to be successful. They also don't tie their ego to their plan.

Four-Star leaders don't have all the answers, but they know how to ask the right questions and work with others to arrive at the answers. Part of asking the right questions is the second contributor to adaptability, which is being able to recognize the current state of a situation, what it

★★

requires, and how it is evolving. Admiral Pat Walsh, paraphrasing hockey great Wayne Gretzky, said a leader must be "able to see where the [...] puck is moving," and move their team there. Things change rapidly; what was going in one direction can suddenly be going in another. We may be on offense one second and defense the next. Adaptability requires being able to redirect—to see that the plan you have begun is not working and not going to work and then being able to go in a different direction. A part of the ability to redirect is being able to overcome the "sunk cost fallacy," our tendency to stick with a failing plan simply because we have already invested so much time, money, and effort into it. Sometimes the best option is to cut your losses, scrap the plan, and go in a different direction. The ability to do that requires the fourth contributor to adaptability: not tying your ego to your plan.

Many of the Four-Stars talked about the role of adaptability, and several provided insightful stories of how their adaptability was vital for their leadership success. For example, General Vince Brooks spoke about being able to recognize the current state of things and where things are going and then being able to redirect as needed. He recalled his own failure to do so during a training exercise while he was leading about 300 soldiers at Hohenfels Training Area in Germany, one that resulted in a rout for him and his men. "[The key is] sensing how the battle is to unfold, the challenge, the task before you is to unfold, and then being agile enough to respond when it isn't the way you thought it would be," he said. "First, to recognize that it isn't. Secondly, to have created options that let you respond when it isn't. Even if it means a 180-degree shift. I was defending, I should've gone on the attack."

When we get locked into our plan and are unable to deviate from it, it can lead to devastating consequences. General Tony Zinni shared how to avoid having that happen by being willing to be persuaded and not tying your plan to your ego.

> Don't fall in love with your plan. [...] Don't fall in love with your own genius. Don't fall in love with what you think is right. Implied in there is, invite others to challenge it—challenge your ideas, challenge your thoughts. Seek from all levels input into what you're doing. Just because you think

★★ ☆ ☆

you've got the most brilliant idea, the most brilliant plan in the world, [that] ain't necessarily so.

Don't tie your plan to your ego

When we make a plan and commit to it, it can be difficult to change directions. As with so many things in leadership, and life in general, there's a lot of psychology involved. As discussed in a prior chapter, we spend a lot of psychological energy trying to maintain our self-image—our ego. We tend to believe we are intelligent, thoughtful, and logical. Because of this, once we have made a plan, our tendency is to do everything we can to make that plan a success, even to the point of following that plan right into failure.

When negative signals come in telling us the plan isn't working, our tendency is to ignore those signals and focus on any and all things that suggest the plan will succeed. This phenomenon is known as confirmation bias, and it affects all of us. Each time that we ignore the contradictory data and further commit to our chosen course of action, we make it increasingly harder to turn away from the plan and go in a different direction. At that point, unwittingly, we have tied our ego to our plan. When we do that, if the plan fails, it becomes likely that we will see it as a reflection of our ability, which translates to seeing it as a personal failure. When we think that way, we often then translate that into a more global assessment of ourselves—"I'm a failure."

While the tendency to tie our plan to our ego has befallen numerous dubious people in leadership positions (e.g., at Enron, Theranos, etc.), it has also stricken scores of great leaders across history. We need only look to Napoleon's march into Russia or Churchill's insistence on attacking Turkey at Gallipoli as examples. This is to say, even great leaders can tie their plan to their ego, which raises a vital question: How can a leader prevent this from happening? Here, through analyzing the Four-Stars' interviews, I was able to formulate four principal ways we, as leaders, can avoid tying our ego to our plan and, thereby, decrease the risk of making tremendous and costly leadership mistakes. These are presented in Table 6.2.

★★

Table 6.2. Ways Leaders Can Avoid Tying Their Plan to Their Ego

WAYS LEADERS CAN AVOID TYING THEIR PLAN TO THEIR EGO
1. Surround yourself with capable people who offer insight and perspective, and allow them to challenge your plans.
2. Allow others to be major contributors to the plan so that it isn't solely yours.
3. Develop multiple contingencies—plan to change the plan.
4. Recognize that success is not the absence of failure. Instead success results from: a. The repeated recognition of minute-to-minute microfailures; b. Constant redirection from those microfailures; and c. In response to those microfailures, continuously reshaping the current plan toward one that will work in the present context and allow you to keep moving toward achieving the goal.

General Eberhart, *on Not Tying Your Position (i.e., Plan) to Your Ego*

As General Ed Eberhart thought about leadership maxims that have guided him, he recalled an admonition he learned from General Colin Powell: "Don't tie your position to your ego." He went on to provide insight into what that meant.

> Essentially, he meant that if you recommend a course of action and your superiors decide that "That's not the course of action that you're going to take. This is a course of action you're going to take," you embrace it, and you run with it. If you tie your ego to your recommendation and they turn [it] down [...], that's a blow to your ego. People have a hard time dealing with that. [...] Now, [...] when you [should] tie your position on an issue to

★★☆☆

your ego [is] if it's an integrity issue, if it's a moral issue, or if it's a life and death issue. Then that's when you fight this to the end, because you can't do something that's not moral, that there's no integrity, or it's a life and death issue that'll cost you lives.

★ ★

CARING

THE THIRD STAR OF LEADERSHIP

CHAPTER 7

CARING:
THE HEART OF LEADERSHIP

The first thing is you've got to care. Caring about troops doesn't mean you're easy on troops. In fact, [...] you're usually pretty tough, and you've got high standards. But troops [...] have this ability to understand whether somebody really cares about them or not and whether they respect them first. I think an exceptional leader's just got to have that. [...] troops [...] want to know that you care first before they worry about your competence. They kind of accept the fact that, in our military, you're going to be competent by the time you get there, but then it's whether you care.

– GENERAL MIKE SCAPARROTTI, U.S. ARMY –

In 2003, then-Major General Marty Dempsey was in Baghdad as the Commander of the U.S. Army's 1st Armored Division. As the insurgents began fighting, his division began taking casualties. As is military custom, for each fallen soldier killed by sniper fire, IED, or other means, there would be a memorial service on the following day at that soldier's forward operating base, and each fallen soldier's footlocker, inverted rifle, helmet, dog tags, and boots would be placed in front of the remaining members of his squad, with all other soldiers from the organization behind them. Along the periphery were the officers and the senior non-commissioned officers

★★★

The chaplain would speak, and the company commander would speak, and then one or two of the teammates of the fallen soldier would speak. It was heart wrenching, but you had to do it to get closure. The last thing that we do is the senior man present files across that line of soldiers and offers condolences and words of encouragement. Then everybody follows this. I went to them all. So, I think I lost in the year 132 soldiers. So, [I went to] 132 memorial services. The first couple of them, I was not very good at it, actually. I mean, I'm being honest with you. They're looking to me to console them and to encourage them, and I just didn't have the words, and I knew I didn't have the words. I would go back and beat myself up about not having the words.

Major General Dempsey would return to his tent at night and wrestle with how he could console his soldiers, how he could show them that he cared deeply for the fallen and for them. One night, in the uncluttered place between waking and dreaming, an answer came to him and jarred him awake.

"Make it matter." [...] I thought, "I don't know what that is, 'Make it matter,'" [...] but it didn't go away [...] We had a memorial service [the next] day, and [I had] this phrase in my head. When the time came for me to go through the line, I reached across and I shook the soldiers' hands, and I looked them in the eye and said, "Make it matter," and it was like a light bulb for me and him. It was like a light bulb because he realized, "Yeah, I can't bring them back." [...] Make it matter, and not just today, but make it matter every darn day. That's what I said. I said, "Just do something every day to make it matter, and in the aggregate of your life, you will have a life that mattered."

General Dempsey found that being at each memorial service and looking into the eyes of each of his soldiers as he stirred them to "Make it matter," conveyed to them how much he cared about them, as well as the fallen. He also found that the mantra made such an impact on him and his soldiers that he had it embossed on the back of the coins he would give out to soldiers—"Make it matter."

Apart from his public admonition to each soldier in the receiving lines at every memorial service, General Dempsey's care for his soldiers spawned a personal practice he has continued every day in the intervening decades.

★★★

I had these cards made up (pulling one from his pocket), and I don't know if you can see that, but it's got a picture of a soldier. It's got the circumstances of how that soldier died. [It's] got something on the back telling about him, where he went to school or [if he was] married, [had] children, whatever it was. I had one card made for every soldier, and I was carrying them around in the combat zone, in my pocket, in my cargo pants. But then I got too many. I told you I ended up with 132. So, I would rotate them. I'd carry three at a time. [...]

I wanted to put [the ones I wasn't carrying] in something special. I just didn't want them lying around on my desk. So, we go downtown [to] a souk (marketplace) in Baghdad. [...] I had somebody go look, and we found a cigar box made out of teak or walnut, [...] and I had something engraved on the cover. [...] "Make it matter." That's where those cards are to this day. Well, I always carry three; 129 in the box. The box's on my desk at home. It was on my desk [when I was the Chairman of the Joint Chiefs of Staff]. It was on my desk [when I was the Chief of Staff of the Army]. It was on my desk at TRADOC. It was on my desk at CENTCOM. Since all this happened, it's on my desk, and that's the phrase, "Make it matter."

Perhaps due to a lack of legitimate experience with the military or because of stereotyped Hollywood representations, some people may be surprised that battle-hardened Four-Stars would discuss caring as one of the principal themes of leadership. However, the best leaders, including those in the military, care about their people. So, despite Hollywood portrayals, it shouldn't come as a surprise that Four-Star officers would care about those they lead. That's been a part of the ethos of our military dating back to the care George Washington showed for his troops at Valley Forge. Across over 250 years of our history, pick an operation, battle, or campaign, and you will find leaders who cared deeply about their troops; leaders who put themselves in harm's way to protect their people and those who died as a result. While it is certain there were those who didn't exemplify caring, they were not exemplary leaders. The best leaders, who demonstrate Four-Star leadership, care about their people.

★ ★ ★

WHAT DOES IT MEAN TO CARE?

There are generally two distinct forms of caring: *caring for* and *caring about*. To care for is directly related to providing for the needs of someone else and may also be referred to as *to take care of*. On the other hand, to *care about* someone centers on socioemotional connection, empathy, and genuine interest in their welfare and happiness. As Admiral Joe Prueher said passionately, Four-Star leaders both care for and about those they lead.

> [...] taking care of your troops, taking care of them physically, taking care of their advancement, taking care of the organization that they're seeing, that they're productive, getting around and giving a damn about what their names are and where they're from and what their interests are. [...] you've got to know your people and you've got to take care of them, and their care and feeding really comes first.

Figure 7.1 presents the components of caring.

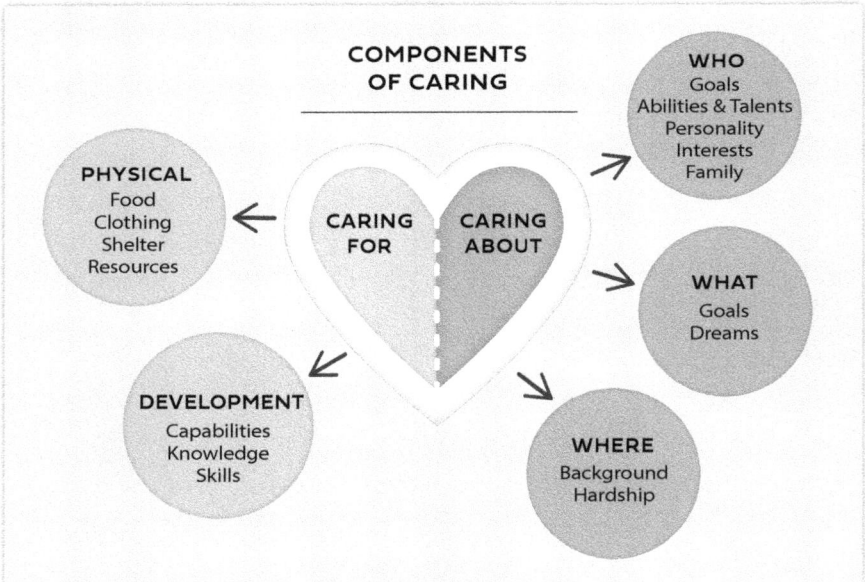

Figure 7.1. Components of Caring. For leaders, caring is comprised of both caring for and caring about the led. Caring for them means providing for their physical and developmental needs, depending on the context of the leadership. Caring about someone means investing in them as a person: who they are, what they hope to achieve, and where they are from.

★★★

Caring for someone is an active process requiring tangible efforts to support their well-being. In the same way a gardener cares for a lawn by providing nutrients, water, and protection from harmful organisms, leaders care for those we lead. Taking care of our people by providing the resources needed to accomplish the task is fundamental to leadership. General Pete Pace concluded, "[...] taking care of your people [is what leadership is] all about. [...] taking care of your people is really critical." To take care of your people involves provision in two domains: physical needs and developmental needs. Depending on the context, providing for the physical needs of those we lead may include supplying food, clothing, shelter, and the resources needed to do their jobs well. Conversely, caring for the developmental needs of our people means we see to it they receive the education and training to ensure they have the knowledge, skills, and capabilities required to perform their jobs well. In these two ways, caring for the people we lead means providing what they require so they can do what needs to be done to accomplish the mission. When we provide for the needs of our people, General Les Lyles said it has direct impacts on mission accomplishment.

> Take care of the people, the people will take care of the mission. [...] if you take care of them, show them that you care as a leader and then understand what their challenges are and try to remedy those, that's going to go a long way to make sure that the mission gets accomplished.

Taking care of those we lead is a fundamental leadership responsibility, one that has been recognized for millennia. For instance, Frederick the Great said, "Whatever beautiful plan you have dreamed up, you will not be able to execute it if your soldiers have not been well fed." Caring for those we lead is a basic duty of leadership.

Whereas *caring for* is related to the physical and developmental needs of someone, *caring about* someone is related to their psychological and emotional requirements—it is tied to their personhood. When you care about someone else, they matter to you as a person. You are interested in them: who they are, where they're from, what they want in life. When you care about someone, you want what is best for

★ ★ ★

them and invest in them as people. However, caring about people can be hard because it takes significant psychological and emotional energy—energy we may not feel like we have. It takes an investment of time to get to know people.

Investing in people psychologically and emotionally can be difficult, time consuming, and, for some, frightening. That investment requires being honest, transparent, and vulnerable, which threatens the image of the omniscient, infallible leader some think they convey. To overcome that takes moral courage like that discussed in Chapter 3. Unfortunately, rather than mustering the moral courage to overcome the risk, many people in leadership positions find it easier and safer to maintain distance, avoid the emotional investment, and present a cold aloofness. What's worse, they usually do so in the name of professionalism, a rather Machiavellian twist. Why should we buy into the lie that choosing to not care about those we lead is professionalism? It seems much closer to egotism, emotional laziness, and psychological insecurity than something palatable or even laudable. The only way to overcome that is to have the moral courage to risk being authentic and vulnerable and make the effort to care.

SELF REFLECTION

What does caring for and about your people look like to you? Would the people you lead say that you care for and about them?

General Janet Wolfenbarger, *on Caring About Your People*

As the Commanding General of Air Force Materiel Command, General Janet Wolfenbarger led 80,000 people and oversaw an annual operating budget of approximately $60 billion. Despite that enormous, high-level responsibility, she has always seen leadership simply, on a ground-level basis: "Take care of your mission, and take care of your

★★★

people." General Wolfenbarger insists a leader's job is "taking care of your people," and that "the workplace must be a place where people feel that they are taken care of." For her, this has always played out in seeing her people as individuals with needs and recognizing that leaders have a responsibility to care for those.

> On the personal side, [...] we as leaders [...] need to understand when our people have stressors in their lives. You need to know your people well enough to be able to understand where people are having stressors and to help them with those agencies that are available or other means by which they can address those stressors. Sometimes, I think there has been a reluctance in the past to feel as though you're overstepping as a leader. If you try hard to help folks on that personal level, [...] you cannot sidestep that responsibility. It's imperative to make sure that your people are in a good place, and that you help them get there if they need [...] Sometimes, I like to think that because of the importance of the mission, because of the hours that we sometimes work in accomplishing that mission in the military, we're with our, I call it Air Force family, more hours of a day sometimes, waking hours, than we are with our real families. So, we are really sometimes in the best position to help folks who are experiencing those stressors to first acknowledge them, to see them, and then, to help them through them.

WHY DOES IT MATTER IF LEADERS CARE?

It is an indictment of our world that there needs to be a justification as to why caring is important to leadership. But since organizations increasingly chase after short-sighted, quarterly numbers instead of long-term success, it is not surprising that they would view their people as short-term, replaceable components that need not be cared for or about. As a result, one of the perennial reasons people leave organizations is because they feel unappreciated—no one cares about them or what they do. Consequently, organizations are facing increasing turnover, financial instability, and growing existential threat—the average lifespan of companies in the S&P 500 has dropped precipitously in recent decades and is now less than 20 years. In the face of these trends, leaders in many newer organizations, along with older ones

★★★

nimble enough to change, are putting great effort into caring for and about their people.

Caring impacts how the leader leads and how the followers follow. Leaders who care consider how decisions will impact those they lead, whether they have the resources and training they need to be successful, and how the individual background, capabilities, and goals inform their roles in accomplishing the mission. This does not mean the leader puts the led above the mission, but it does mean the leader is more thoughtful about the mission and how each person fits into achieving success, which invariably increases the likelihood of success. This balance can be difficult to strike, and many leaders come down on one side or the other. Leaders fall short by either focusing solely on the mission to the detriment of those they lead or, conversely, concentrating on caring for and about their people at the expense of mission accomplishment. To accomplish this balance, leaders need to be emotionally invested without being emotionally attached. That is, as leaders we need to care deeply about our people and invest in their success and well-being, while also maintaining enough emotional distance to remain objective and be able to make the difficult decisions when necessary.

Followers who know their leaders care for and about them are more committed to the leader and the mission. Numerous scientific studies, including a large 2002 analysis by Kurt Dirks and Donald Ferrin of data from over 100 independent studies, confirm what we intuitively know; when people believe their leader cares for and about them, they trust their leader. So far, investigators have not been able to numerically quantify the relationship between the magnitude of a leader's caring and the degree of followers' trust. However, Figure 7.2 presents a reasonable conceptualization of that relationship. For most of us, someone must demonstrate they care a fair amount before we begin to trust them. However, once they have demonstrated a higher level of care, our trust in them increases rapidly. Because of this tendency, I have represented the relationship as non-linear.

Beyond increasing trust, when people know their leader cares, individual engagement and team morale increase. The converse of these

★ ★ ★

things is that when a leader does not care about their people, the people won't care about the mission, and that all but guarantees failure. However, while it is clear we must care for and about our people, as General Chuck Wald cautioned, a leader's caring must be sincere. "You can't fake it," he said. "If [the people you're leading] don't believe you care about them, [...] then you eventually will fail one way or the other. Either they'll fail you, or you'll fail the mission because you can't get them to do what [you need or] want them to do."

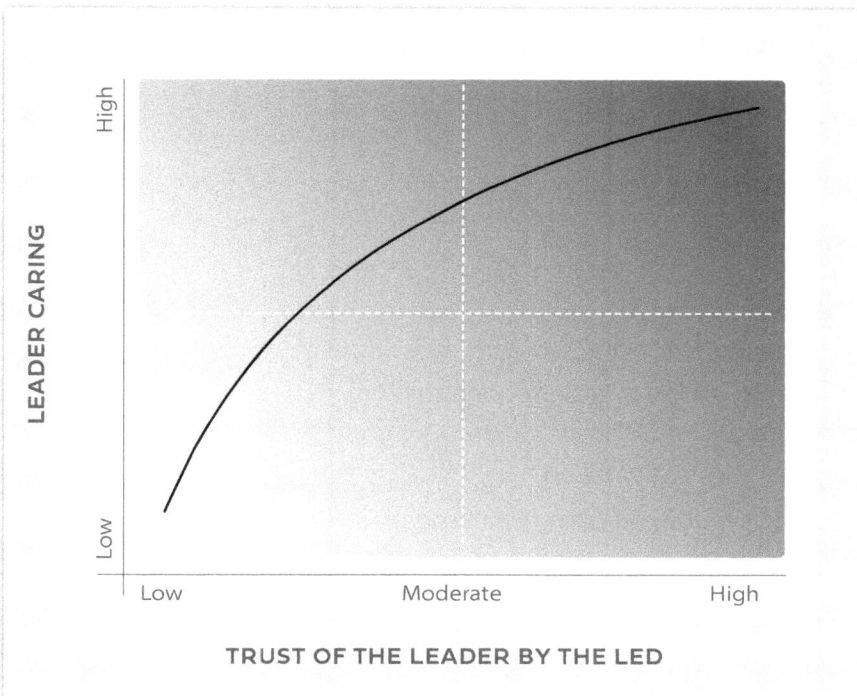

Figure 7.2. Conceptualization of Relationship Between Leader Caring and the Level of Trust the Led Have in the Leader. As the leader's degree of caring increases, trust amongst the led increases. A significant degree of caring must be present before the led begin to demonstrate a high level of trust.

The remaining chapters in this section concentrate on what it looks like to focus on and care about those we lead.

★ ★ ★

CHAPTER 8

IT'S NOT ABOUT YOU

It's a privilege to be placed in a leadership position, and it's not about you. It's about doing right by your mission and by your people.

– GENERAL JANET WOLFENBARGER, U.S. AIR FORCE –

An Army division is typically comprised of 10,000 to 15,000 soldiers divided amongst five to seven brigades, depending on its mission. When he was a brigade commander, General Gus Perna was disappointed to find some of his fellow commanders, who should've been putting their mission and people first, were focused on themselves.

> It used to drive me crazy to go into a meeting with seven brigade commanders, not all of them, but inevitably there was one or two that thought that they were the best and the most important brigade commander and brigade in that division. They were the ones that were always [whining], "I need more money. I need more training time. I need more range time. I need more love."

Those colonels did not understand or care that the division commander had a fixed amount of resources that must be distributed in such a way to assure the entire division was trained and ready to fight. Instead, in General Perna's assessment, the one or two colonels' egos would overshadow the common mission, and they would advocate only for themselves, telling him, "I'm the best, I need more, I'm smarter. [...] If you don't pick me to go on this mission, then I won't be recognized

★★★

[...]" In those situations, General Perna had to be stern to help his sub-ordinate leaders understand their responsibility to their role and the greater mission. That included taking their turn being the brigade to go around and clean-up cigarette butts and other trash while their counterparts were training.

> Hey, if you get told to go do Red Cycle, [...] you're going to go do all the ash and trash details around the division so two other brigades can train. Well [colonel], go be the best at your ash and trash, be accountable to it, be right to it. That's thinking bigger than yourself. So, the other two brigades can go freaking be ready to go, and you'll get your turn when you get your turn, and you'll get your money when [I] as a division commander see fit.

In General Perna's judgment, part of the reason some leaders struggle to put the mission and others first is because most hierarchical leadership structures are predicated on individual competition—an individual must be better than others to get promoted into a leadership role. That process can breed a "me-first" orientation in many people. General Perna countered this by encouraging his people to embrace the reality that leadership is not about looking out for yourself—he helped them see how failing to recognize that can be disastrous.

> What I would tell people is, "Do your job; be good at your job, but we're a team sport. If you're in a perimeter and your perimeter is good—you did everything right to standard—but the perimeter on the left and on the right of you sucks, you're going to die." It's really pretty simple: if your wing man sucks, you're going to die.

Most people in leadership positions have arrived there because they have performed well as individuals. They have spent years, likely most of their lives, focused on their own skills, achievements, and success. So, it probably shouldn't be surprising that when they are placed in a leadership role, many continue to do what they've always done—focus on themselves. From a psychological perspective, that makes sense. It is operant conditioning—we get the behavior we reward. If a person has excelled as an individual and is, in some respects, rewarded for such an orientation with a leadership position, we can expect that

★ ★ ★

same behavior to continue. In his book, *On Leadership*, John Gardner identified the tendency for leaders to continue to focus on individual performance:

> Most young people in professional and executive ranks have had long training—literally since elementary school—in individual performance. They learn that it is how they perform as individuals that counts, not how they relate to others. So, it is not surprising that many young executives— even middle-aged executives—are still pirouetting for some scorekeeper, real or imagined, with little thought of their possible constituency. (p.167)

When someone has spent most of their life focused on themselves and their personal success, they are likely to lead in a similar manner—focused on their ideas, goals, opportunities, accomplishments, and rewards. They see their leadership as being about serving themselves.

But leaders who are out to serve themselves are their own worst enemy, undermining their own success. Those they are attempting to lead recognize the egocentrism and disengage, which leads to leadership failure and career disappointment. General Jim Jones cautioned against egotism saying, "[...] understand that whatever happens is not about you." During his career, he observed that some people did not understand that principle and its impact, saying, "[...] those that seemed to be more concerned about their own climb up the ladder and what was going to happen to them didn't make it very far." General Keith Alexander saw the same thing in the careers of egocentric leaders.

> [...] the guys who were kind of the peacocks, [...] the ones who were out there, "See how good I am, see what I'm doing," their hair was combed exactly right. They stood up, and they would talk about "what I did" instead of "what we did." They [...] only went to maybe the One [or] Two-Star level, and [...] were moved out. It's really good, but we're looking for somebody who can build the team.

Four-Star leaders understand that they have an important role, but also know that leadership isn't about them. General Vince Brooks said, "Don't lose sight of your importance as a leader, but don't get so focused on [it] that you think that it's about you. It's not." Instead, he reiterated, "It's about the led." As a leader, always remember it's about the mission

★ ★ ★

and those you lead, not about you. Without you, another leader will be found, and success can be had. Without those you are trying to lead, you are only a solitary voice with an ill-fated idea and no hope of accomplishing the mission. If you hope to be effective as a leader, your focus must be on those you lead. Figure 8.1 presents the relationships between leadership effectiveness, the leader's egocentrism, and the leader's focus on those they lead.

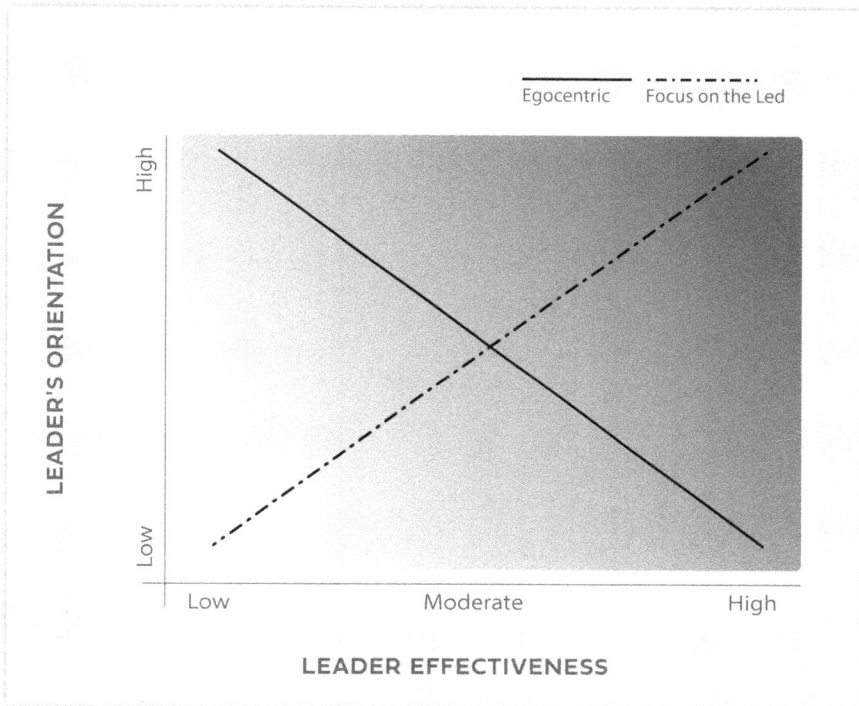

Figure 8.1. **Conceptualization of the Relationship of Leadership Effectiveness with the Leader's Level of Egocentrism Versus a Focus on the Led.** Generally, leaders who are egocentric have low leadership effectiveness. The more a leader focuses on those they lead, the more effective the leader will be.

In addition to bridling an individual orientation, it is also important to recognize that the mission is bigger than your role in it. The brigade commanders General Perna discussed are illustrative of leaders who were primarily focused on themselves, and when the mission served their personal objectives, their immediate personnel. The brigade

★ ★ ★

commanders' reluctance to take their turn as the clean-up crew while the other two brigades trained showed their willingness to sacrifice both the division's overall good and the accomplishment of the larger mission. That must not happen. In today's organizations, success hinges on leaders who prioritize our teams' achievements within the context of the greater mission.

If we accept that leadership is not about the leader, it may be difficult for those leaders who have always performed as individuals to understand what it looks like to focus on those they lead. The Four-Stars recommended numerous ways to help leaders shift to that focus. Three of those will be covered in detail in the remainder of the chapter.

WHO IS MOST IMPORTANT IN A LEADER'S EYES?

As a young lieutenant serving in Vietnam as a platoon leader in the 101st Airborne Division, Tom Hill had a particular soldier in his platoon who recognized that then-Lieutenant Hill cared about his men. As a result, for decades after, when he had no one else to call, that soldier would call General Hill.

> [...] from Vietnam until he died, about every eight [to] 10 months, I'd get a phone call [at] two or three o'clock in the morning, and I'd have to talk this guy off a ledge. He would've fallen off the wagon. One time, I literally talked him off a ledge, and I made a difference in his life. I kept him alive for a long time or helped keep him alive for a long time. That's what you've got to do. Make a meaningful... That's what I want to be remembered [for]: I made a meaningful difference in someone else's life.

Chapter 3 covered the idea of service over self within the context of character, the first of the four stars of leadership. It focused on commitment and dedication to the mission, people, and organization as a manifestation of character. In this chapter, a leader who puts other people over themselves is focused on caring for and meeting the needs of those they lead before tending to their own needs. As discussed in the prior chapter, that includes not only their physical needs, but also their developmental needs. General Lori Robinson recalled when she began to understand that her job as a leader was to put her people and their needs over her own, and that realization informed what she did for the rest of her career.

> [...] when I became the Chief of Tactics and I had people working for me, really for the first time ever, it started to dawn on me that it really actually wasn't about me; that it was actually about everybody else. In that role, my goal was to have everybody that worked for me in Chief of Tactics go through and graduate from the Fighter Weapons School, and they did. My goal was to make them better than me. In fact, a couple of the guys got promoted early to lieutenant colonel. I never got promoted early until colonel. So, it made me understand that what's the most important thing is making those around you better...

Thereafter, when General Robinson took on a new position, she let everyone know her leadership philosophy and her goal to put them first. "Every time I took command, I'd bring my subordinate commanders in and say, 'My job is to make you a better squadron commander than me, group commander than me, then wing commander than me,'" she said.

Servant leadership is a well-known leadership philosophy covered in countless books. At its core, servant leadership is about prioritizing the needs of those we lead over our own needs and interests and then acting on that prioritization. General Charles Krulak said as a leader, "you care more for the people around you than you care for yourself." That doesn't mean that you don't care about yourself or tend to your own needs. Rather, it is in the prioritization—those you lead must come first. This was a common theme critical to leadership among the Four-Stars. For instance, General Vince Brooks defined leadership in this way:

> [As a leader, you must always remember] that your leadership is fundamentally about the led. It is not about your position. It's not about your attainments. It's about the led—those who are led by you at any given point in time. If you keep that mindset, then you'll have a servant leader's heart where your purpose is them: making them better, making them more effective, making them survive in combat if that happens to be military leadership, or achieving the bottom line better, feeling better about themselves, staying emotionally and mentally healthy. It's about them.

★ ★ ★

WHO DO YOU PUT FIRST WHEN YOU'RE UNDER ATTACK?

In 1968, the war in Vietnam was raging, and then-Captain Barry McCaffrey was made a company commander in the 7th Cavalry after the prior commander was killed in combat. Three days after assuming command, McCaffrey was given orders that his company was going to be flown into a firefight to rescue another company being overrun by the Vietcong. He gathered the company's leaders, explained the situation to them, and, with his hands shaking, gave them the order to prepare to fly out. Before departing, he told his first sergeant, "You be in the last helicopter; I'm going to be in the first one." It was after dark when the fleet of 16 Huey helicopters was approaching the landing zone.

> We come in, and we're under groundfire. My helicopter got hit with a loud "Bam!" right up through the floor, and my 18-year-old radio operator, [...] [with a] big smile on his face, gives me a thumbs up. We'd just been hit by a 51-caliber machine gun. We get on the ground. There's incoming mortar fire. It's now dark; tracers whipping through the night. I got them organized. I told them, "At 12 o'clock, I'll give you the final order [to move off the landing zone]."

Shortly thereafter, he gave the order to move out into the darkness. As the company moved out, they were holding onto each other's equipment to stay together, and the person behind McCaffrey kept banging into him. McCaffrey became annoyed and turned around to address the soldier who kept smashing into him. He was surprised by what he saw.

> I turned round. It's a big man. He's got a box of machine gun ammo on his shoulder. It's the battalion commander, Tad Davis (McCaffrey's superior). He didn't say one word to anybody. He just got on one of the helicopters. I asked him, "Sir, why didn't you speak up?" "Well," he said, [...] "It looked to me like you knew what you were doing. Why should I have spoken up?"

The leadership mindset that "it's not about you" is, perhaps, nowhere more clearly demonstrated than in the U.S. Marine Corps' ethos of "Officers eat last." Originally born in the Marine Corps out of necessity and as a demonstration of caring for troops, this principle has since been disseminated across all branches of the U.S. Military. It was also

★★★☆

popularized as the subject of Simon Sinek's book, *Leaders Eat Last.* Consequently, the concept has made it into the mainstream, and leaders outside of the military are becoming more familiar with it. Yet, it is clear in corporations, government, and society in general, many in leadership positions either haven't yet heard of the idea or, perhaps out of persistent egocentrism, they have expressly chosen not to practice it. Perhaps because of this, it was a concept multiple Four-Stars spoke about.

"Officers eat last" represents a combination of two principles of excellent leadership we have already covered individually: caring for those you lead and maintaining personal humility. Recall that Four-Star leadership requires a commitment to supporting the welfare of those you lead. It also demands not regarding yourself as better than anyone else. Again, that is not to say that leaders should disregard ourselves or our own personal welfare; rather, it is prioritizing others first, as discussed earlier. When leaders ensure that people have what they need to do their jobs—and feel *cared for*—they will also believe they are *cared about*. Caring fosters the trust that empowers leadership, resulting in greater individual and organizational mission accomplishment.

Caring for and about those you lead is both an ethical and practical imperative for leadership exemplified in the idea that "Officers eat last." General Jim Jones, who was the Commandant of the Marine Corps and went on to serve as the Supreme Allied Commander Europe, recalled learning the concept of "Officers eat last."

> My father, a WWII Force Recon Marine officer, used to say that in the Marine Corps, "Officers eat last." I used to ask him why that was. He explained, "Because the officers have to order the food and in order to make sure they order enough food to feed everybody, the only way to find that out is if the officers ate last." If they ran out of food, it's the officers that wouldn't eat, not the troops. [...] You don't take care of yourself first and make sure you're comfortable and you've got everything you need before you make sure that the men have everything they need.

As alluded to by General Jones and his father, "Officers eat last" is about the leader's willingness to prioritize the needs of those we lead over our own, whether that is regarding food, shelter, supplies, etc.,

★ ★ ★

and, like General Jones, multiple Four-Stars were clear that putting first those you lead is "the right thing to do." Beyond the ethical component, however, there are also significant practical implications of "Officers eat last." For example, General Glenn Walters highlighted the practicality of it.

> [...] leaders eat last. That's to make sure the troops get fed first, in case they're short. But there's a pragmatic part to it too, because if you're doing it right, they're doing all the heavy lifting; you're doing the heavy thinking. They're probably going to need the protein and the energy more than you. That's why you want them to eat first.

General James Conway, who saw his share of combat action and went on to serve as Commandant of the Marine Corps, added to the practical reasons that officers eat last.

> [...] the officers eat last [...] When you get hot chow brought to the field or you're in combat and we have hot chow out there, morale goes up 10 points on every occasion. But, when that formation lines up, it's the (lowest ranking) private that's about to get booted out of the Marine Corps in three weeks who's the first guy in line, and the last guys in line are the battalion commander, or whatever level commander (highest ranking officer), and the sergeant major (highest ranking enlisted Marine), and you line up that accordingly in between. The thought process is every man in that unit is valuable. The more junior, the more valuable because he's going to be closing the last 300 meters under machine gunfire. And you want him or her [...] to be provided for as best you possibly can. The meal in the field is small potatoes, but the mentality, the symbolism is what carries through.

Not only does "Officers eat last" ensure that those on the frontlines have what they need, but it also demonstrates that they are highly valued. In most cultures, when a meal or banquet is held, the guests of honor are served first. As a result, when we leaders choose to eat last, it is an active demonstration of respect and honor being given to those we lead, and they cannot help but feel cared for and valued. Additionally, most people will recognize that while it does not necessarily mean we leaders will not eat, it may well mean there is less, the palatable choices are

★ ★ ★

gone, or the food is cold. Whatever the case, "Officers eat last" clearly demonstrates caring for the led, and they recognize and appreciate it.

"Officers eat last" is not merely one more in a litany of prescribed leadership actions but rather is an ethos, a fundamental guiding principle. This philosophy requires us to prioritize the well-being and needs of the people we lead over our own. It also serves as a lens through which we leaders should view ourselves and our role, fostering a perspective of personal humility. While increased efficiency and productivity may be positive outcomes, the essence of "Officers eat last" lies in its emphasis on selfless leadership and a genuine concern for the welfare of those we lead.

WHAT DO PEOPLE NEED?

Upon retiring from his post as Commander-in-Chief of the U.S. Pacific Command, Admiral Joe Prueher was appointed to be the 7th U.S. Ambassador to China. Arriving in Beijing, he was disappointed to see the state of the U.S. Embassy and the living accommodations of some of the State Department's junior personnel.

> When I got to the Embassy, which was quite old, it was apparent, there was a big discrepancy in living conditions for State Department people. Beijing was modernizing from the 1980s, and about 25% of our people and their families were in suitable apartments. Living conditions for many were badly substandard. Several families were in a Soviet-era high-rise where in the worst case, a few other nation's diplomats sacrificed goats in the vestibule! Obviously, from a military background this situation was intolerable. Yet, when the State Department sent visitors to see us, these visitors were booked in a nearby St. Regis Hotel, and heretofore were not exposed to the underserved side of housing.

Admiral Prueher was dissatisfied to see the living conditions of the junior personnel, and he sent a message to Secretary of State, Madeline Albright, a communication known in the Navy as a "UNODIR"—which stands for "unless otherwise directed." It was a way for him, as the subordinate, to inform her (as his boss) of his intentions and provide a tactful opportunity for her to override his plan. He let her know that

★ ★ ★

moving forward, unless she directed otherwise, the accommodations would be changing for all visitors from the State Department.

> [I advised her], all visitors from the State Department would stay in the high-rise housing rather than the St. Regis. Further, we planned to let anyone who wanted to have the choice of moving to a sufficient apartment in Beijing. New arrivals would be appropriately housed. I never got a response to that message. So, we did it.

The foreign service personnel in the Embassy were unaccustomed to the military notion of "take care of your troops" or the idea that someone might advocate on their behalf with the Secretary of State. Admiral Prueher's commitment to the well-being of his personnel had far-reaching effects on the morale, *esprit de corps*, and performance of the organization.

> Just the fact that we made that move to "take care of your troops" had unforeseen benefits. That, plus the example set by a handful of star performers at various levels made it seem like the cadence of the band in the Embassy had picked up. Some really talented foreign service stalwarts commenced to work near peak performance; people even dressed better; energy level was higher. When some real challenges with the People's Republic of China in the national security realm arose, our team was, in fact, a team which achieved some real breakthroughs.

Recall from Chapter 7 that there are two components of caring for people: physical needs and developmental needs. Whereas "Officers eat last" is focused primarily on assuring the physical needs of your people are met before your own, supporting your people focuses more on ensuring their developmental needs are met. In the foregoing example from Admiral Prueher, while he supported his people by improving their housing, his support proved to be particularly important in developing an environment of camaraderie and performance that allowed the organization to weather difficult circumstances.

There are two principal components to supporting your people. The first facet of our support of our people is ensuring they have those things needed to do their jobs well. This is, in part, *caring for* people. It would be ridiculous for a military leader to assign a bomber pilot to fly a mission to attack a target while only providing the pilot with

★★★

a bicycle. Yet, something similar happens in countless settings every day when leaders insist that their people make do in the absence of appropriate supplies, equipment, or personnel. Organizations across nearly all domains fall prey to this by insisting on continued use of archaic software programs. Healthcare organizations persist with inefficient electronic medical records and outdated medical equipment. Manufacturers continue using dilapidated machines. Usually, these sorts of "make do" practices are framed as financial stewardship, in a manner not too dissimilar from how "work force reduction" is often heralded as an exercise in efficiency and value. At its heart, it is not providing the support the people need to do their jobs well.

Four-Star leadership demands caring for the functional needs of those we lead so they can focus on and accomplish their mission. To be able to do that, we must know what they actually need to get the job done, which means understanding what is required of anyone to accomplish a given task. While there clearly are contextual specifics, I have identified at least six requirements any given person needs if they are to accomplish anything.

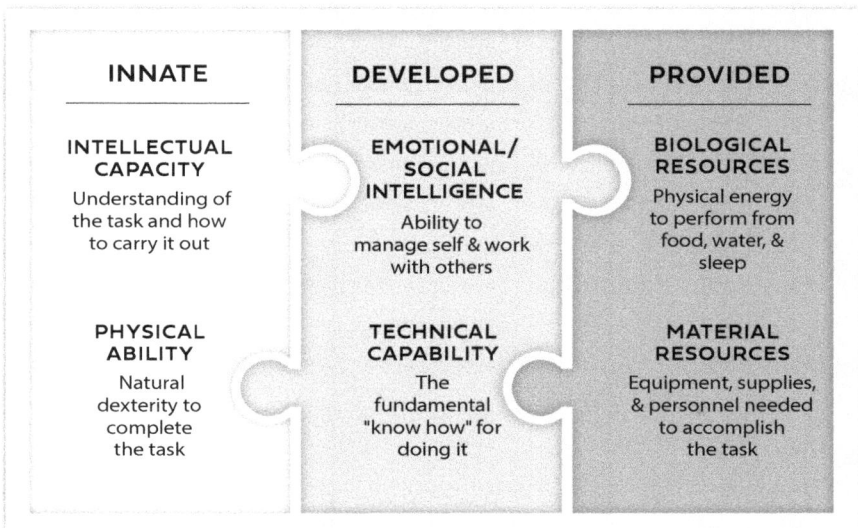

INNATE	DEVELOPED	PROVIDED
INTELLECTUAL CAPACITY Understanding of the task and how to carry it out	**EMOTIONAL/ SOCIAL INTELLIGENCE** Ability to manage self & work with others	**BIOLOGICAL RESOURCES** Physical energy to perform from food, water, & sleep
PHYSICAL ABILITY Natural dexterity to complete the task	**TECHNICAL CAPABILITY** The fundamental "know how" for doing it	**MATERIAL RESOURCES** Equipment, supplies, & personnel needed to accomplish the task

Figure 8.2. Requirements for Task Accomplishment. The six requirements needed for the accomplishment of a given task fall into three categories: those that are innate; those that can be acquired through development or training; and those that must be provided. As shown, the three categories interlock with each other, indicating their interdependence.

★ ★ ★

Grasp the requirements for accomplishing a task

Figure 8.2 presents the model I have developed for those things need-ed to accomplish a particular task. In it, there are three categories into which the six requirements are distributed: innate, developed, and provided. With two requirements each, the categories are interde-pendent—a task won't be accomplished if one of the items is missing.

Intellectual capacity and physical ability are inherent to a given person. The intellectual capacity for understanding complex topics—someone's intelligence—is largely innate, though work by investiga-tors like Carol Dweck has shown it is not static and can be developed. If we consider what is required to fly a fighter jet, the pilot must have the intellectual capacity to read and very rapidly assimilate all the data from the cockpit instruments. The pilot also must be able to identify threats and targets, determine the appropriate action to take against them, and act on that decision nearly instantaneously. Would-be pilots with congenital or acquired cognitive difficulties may not be able to accomplish some of those things. Similarly, for any given task, there is some degree of physical ability required. Continuing with the example of the fighter pilot, a blind person does not have the visual ability to fly a fighter jet and, therefore, could not be a pilot. Consequently, as a leader, you must make certain the person chosen for a given task or role has the physical and intellectual abilities to accomplish it. If you don't, they will fail, and the task will not be accomplished. For a fight-er pilot, this is assured through regulations; there are requirements for height, weight, visual acuity, and a host of other things. There are also regulations for physical fitness, as well as test and training scores. The same is true for other high-performance fields. If we want to lead high-performing teams that accomplish the mission, we have to iden-tify the physical and intellectual requirements and match our people to them.

The second category of requirements for completing a given task is comprised of those things that can be trained or developed. Here, the two requirements are technical or tactical capability and emotional/social intelligence. Technical capability is the fundamental "know how"

★ ★ ★

to do something. A person may have the physical ability to accomplish something, but that does not mean they actually understand how. Returning to the example of the fighter pilot, a person may have all the intellectual capacity and physical ability to fly a fighter jet, but until that person is trained, they won't be able to. The second requirement that can be developed is emotional/social intelligence. As described by Daniel Goleman in his influential book by the same name, emotional intelligence regards a person's ability to recognize, understand, and manage their emotions, as well as perceive, interpret, and respond to the emotions of others. This is accomplished through self-awareness, self-regulation, social awareness, and relationship management. In our ongoing example, fighter pilots must be able to manage their emotions, such as fear, anxiety, or anger, so they can fly and maneuver their jets in highly stressful circumstances. Additionally, because fighter pilots do not operate in isolation, they must be able to effectively manage relationships with others. Both technical proficiency and emotional intelligence are essential things that can and should be developed in those we lead. It is incumbent on the leader to recognize this and ensure their people receive training to enhance both. Doing so fosters mission success, and failing to do so promotes mission failure.

The final group of requirements for the accomplishment of a task are those provided by the leader, both biological resources and material resources. Biological resources are those basic physiological needs a person must have to ensure the physical energy needed to accomplish the mission: food, water, shelter, and sleep. It could be argued that, unlike the military, in most organizations it is not the role of the leader to provide those biological needs. While true in a direct sense, it is not true in a larger functional sense. Fundamentally, if someone is employed by an organization, it is incumbent on that organization to compensate the person adequately for their service so they may be able to have adequate food, water, and shelter. If we leaders do not make certain our people have those things, we can expect them to lack the energy needed to accomplish their task. Similarly, humans have a fundamental need for sleep, and without it, we do not function well. Therefore, it is in our best interest to make sure our people get enough

★ ★ ★

sleep, which means ensuring they are not working too many hours per day or week. Few people would advocate for our fighter pilot exemplar to get into the cockpit without having had adequate sleep. Who would want to ride along with that fighter pilot? It is this idea that has driven drastic changes to the work-hours of physicians-in-training. When I began my medical residency, there were no work-hour restrictions, and I worked in excess of 120 hours per week. The Accreditation Council for Graduate Medical Education recognized this could present a problem for patients and trainees and established the 80-hour rule instead.

Apart from the biological resources needed to accomplish a task, the final requirement is for material resources. The fighter pilot must have a jet with appropriate avionics, ammunition, fuel, and a runway, not to mention personnel to maintenance and fuel the jet, load ammunition, and direct takeoff and landing. This returns us to the requirement for leaders to understand what is needed by people to accomplish the mission. If we cannot provide the resources to accomplish the mission, then we need to seriously consider redirecting to a different mission. As leaders, if we have a proverbial sow's ear and want our people to make a silk purse from it, we are going to end up with a sow's ear purse. We have a responsibility to make sure those we lead have the materials required to do the job to the level of excellence required.

Be an encourager

While ensuring people have what they need to do the job is the first component of supporting people, the second component is encouraging them. This was a common theme during the interviews and for good reasons. While we all know from our own experiences that we feel more confident and have a better outlook when someone encourages us, the positive effects of encouragement extend beyond our individual feelings and outlook and impact the teams we are a part of. Numerous scientific studies have found highly valuable effects associated with leaders who are positive and encouraging. These effects are related to both individuals and the teams they lead. Figure 8.3 summarizes these findings.

★★★

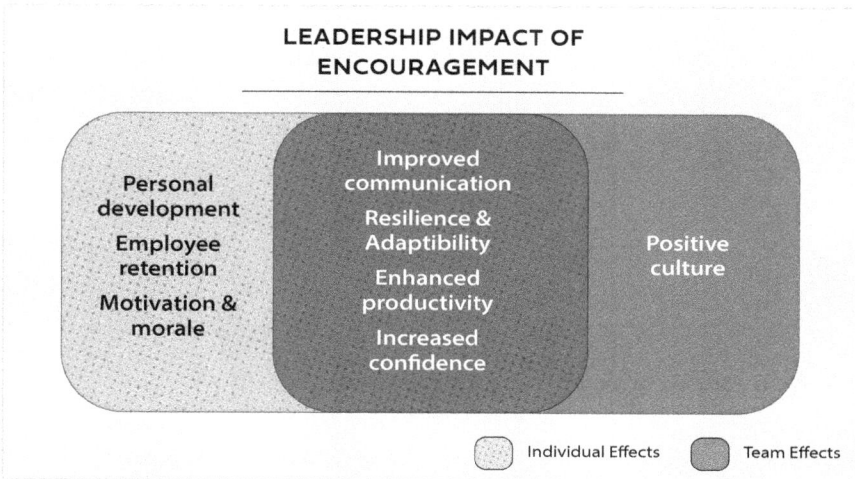

**LEADERSHIP IMPACT OF
ENCOURAGEMENT**

Personal development
Employee retention
Motivation & morale

Improved communication
Resilience & Adaptibility
Enhanced productivity
Increased confidence

Positive culture

Individual Effects   Team Effects

Figure 8.3. Impact of Leaders Who Encourage. When leaders encourage their people, their encouragement has powerful individual effects, as well as crucial effects on the entire team.

The powerful effects of encouragement go a long way toward helping individuals and teams accomplish their missions, which means encouragement is instrumental in effective leadership. General Skip Sharp endorsed the importance of this when he expressed that he wants to be remembered as "a leader who encouraged and built teams to be able to accomplish the mission."

For Four-Star leaders, encouragement is more than simply positive affirmation. It is seeing the potential in people and working to develop it. In fact, a hallmark of Four-Star leadership is the development of subsequent leaders and, consequently, the great things they achieve. Encouraging and supporting the development of those they led was particularly important and meaningful to multiple Four-Stars. For instance, Admiral Prueher said that leadership is "about [making certain the people you lead] get opportunities to advance." This was of the utmost significance to his view of leadership, to the degree that he hopes his leadership legacy would be "to be remembered by what [his] people did, the successes they had." General Gus Perna also spoke about his hope for his legacy to be that of supporting and developing his people to do great things.

★★★

> What I want to be [remembered for is] the people that are successful after me, that I trained. I coached them, and taught them, and mentored them, and gave them opportunities to succeed and fail. Not only are they successful, but they went way beyond what I was. That's what I want to be remembered for, to be honest with you. I think that's the biggest thing that we can do in our profession...

General James Conway felt similarly, saying, "What's important is that you [...] develop your people to the best of your ability," adding, "your accolades [as a leader] come when people realize [how great your team is]."

Ultimately, Four-Star leaders recognize that it's not about themselves—it is about those they are privileged to lead. Consequently, they put the led first, making certain the physical and developmental needs for mission accomplishment are met. Beyond that, Four-Star leaders understand that their legacy as leaders will not be judged by the personal accomplishments that got them to their positions, but rather their leadership legacy will be lived out in the lives—successes and failures—of those they led and developed. General Stanley McChrystal summarized this well.

> It's always about the people, and it's so easy to forget that. But I say this, not just because you want to be thought of as a people-person leader, but the way to be successful as an organization is through the people. If you take care of people, they have cohesion, and they're professionally developed, the rest just works. If you don't do that, the rest does not work.

SELF REFLECTION

What are you actively doing to support the development of each person whom you are responsible to lead? What is your leadership legacy that they will live out?

★★★

CHAPTER 9

CARE ABOUT YOUR PEOPLE

Your subordinates care that you care about them.

– GENERAL PETE PACE, U.S. MARINE CORPS –

General J.D. Thurman served as the Commander of the Army's 4th Infantry Division in Iraq from 2004 until 2007 and cared deeply about his soldiers. During his interview, over 15 years after leaving Iraq, with somber weight, he spontaneously recited his division's casualty numbers.

> I can tell you every day what happened in Iraq. I lost 235 soldiers, had 1,354 wounded, 47 amputees, 25 double or triple [amputations]. To this day, I can tell you where all those people were killed. I called their mother or father or the spouse of every soldier I lost. To this day, I feel accountable for that, and I went to every memorial service and all that. When you lose a soldier, it's just... That's what goes on out there. The soldiers, men and women seeing that [say], "Hey, my commander is here. He cares." That carries a lot of weight. It carries a lot of weight.

While caring for the needs of those you lead takes operational effort and, sometimes, personal sacrifice to ensure their physical and developmental needs are met, caring about people requires a great deal more personal investment on our part as leaders. We can systematize and automate *caring for* people's physical needs and training, but we can't automate *caring about* someone. That requires a personal emotional investment in each individual and their wellbeing, which

★ ★ ★

demands a significant vulnerability on the part of the leader. It means we are vulnerable to emotional and psychological injury if something bad happens to someone we lead or if that person chooses to leave the group or organization. That degree of personal investment and vulnerability on the part of leaders is particularly stark in the military profession where death is, for some, an unfortunate outcome. It is also, undoubtedly, why so many of the Four-Stars could, after intervening decades, still recount from memory the time, place, and names of those they had lost—they cared deeply about their people. Beyond the military and public service, such grave losses aren't expected in most professional fields. Yet, that doesn't mean there isn't a risk of substantial emotional injury. Terrible things happen to people in all walks of life, in fact, much more frequently than they happen to those in harm's way. Brain tumors develop, cars crash, and houses burn down, sometimes with people in them. Everyone faces tremendously challenging situations and the risk of losing those we care about. Four-Star leaders confront that risk head-on.

Caring about those we lead is at the heart of Four-Star leadership and was portrayed in Chapter 7 as one of the two major components of caring: *caring for* and *caring about*. Caring about someone is so importance to developing Four-Star leadership that it warrants this chapter dedicated to it. Here we will cover the three principal components of caring about those we lead. These are presented in Figure 9.1. These three components shouldn't be seen as prescriptive, but rather are reflective. That is, they reflect those fundamental facets of someone and their story that we all find important about those we care about.

As General Thurman discussed caring about his soldiers, he said, "People know if you care about them." Think about your own life. How do you know when someone cares about you? For almost everyone, it is that we feel, on a fundamental level, that we are cared about as a person, an individual with a personality, a background story, and hopes and dreams. We aren't a number to be reached, position to be filled, or obligation to be endured. In those moments when we know someone cares about us, we feel seen and appreciated as an individual.

★★★

FACETS OF CARING ABOUT SOMEONE

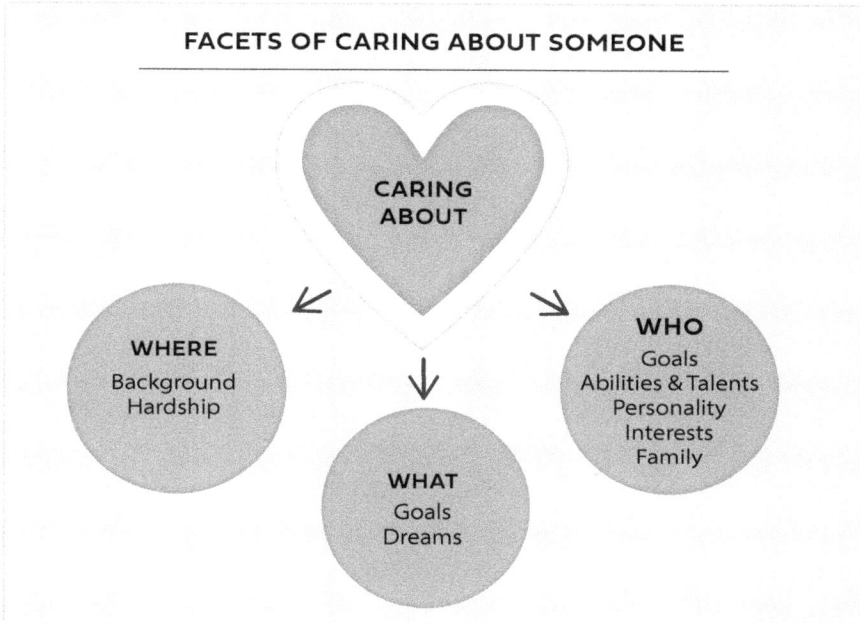

CARING ABOUT

WHERE
Background
Hardship

WHO
Goals
Abilities & Talents
Personality
Interests
Family

WHAT
Goals
Dreams

Figure 9.1. Facets of Caring About Someone. Caring about someone includes at least three major facets: who they are as an individual, where they come from, and what their hopes and dreams are.

There's more to those moments than feeling seen and appreciated. When we feel seen and appreciated, our brains produce oxytocin, which increases our positive feelings for the person who has cared about us. The boost in oxytocin and positive feelings increases feelings of trust, which strengthen our relationship. Consequently, it should come as no surprise that numerous studies have shown positive and beneficial effects when leaders care about those they lead. Included in those are enhanced job performance, positive team dynamics, and organizational commitment, as well as increased well-being, job satisfaction, and engagement. You must have those among the people you lead if you are to ever be successful as a leader.

General Tony Zinni identified caring about people as one of the three most important leadership keys he learned in his career, saying. "Leadership is about people. So, it's how you relate to people, how you demonstrate your sincerity in caring about their welfare and who they

★★★

are." He then told a story of how caring about someone—who they are—can have a profound impact.

> When I was a captain and company commander, my company was leaving on a deployment, and we had just pulled out of Morehead City, North Carolina, headed to the Mediterranean, and I came out on deck. It was evening, and there was one of my Marines, a really outstanding young African American Marine. I knew all about him because we had just promoted him to meritorious corporal. He had come from an inner city and never had a father—really tough life—but [had] joined the Marine Corps and was exceptionally great. I was telling him how proud I was of him and everything he had done and what he had accomplished. He looked at me, and he said, "You know, sir, you're the closest thing I've ever had to a father." It kind of took me back on my heels.

Caring about people as individuals is like fertilizer for growing strong relationships; without it, relationships can develop, but they aren't as strong, healthy, and productive. Four-Star leaders know this, but they don't exploit it. They care about their people because it is the right thing to do, not out of a Machiavellian deceit aimed at enhanced leadership productivity. Remember the words of General Thurman: "People know if you care about them." Few things can have worse effects on your ability to lead than feigned care revealed to be a manipulative charade.

Within caring, being kind and showing empathy to those we lead are important factors for successful leadership. The three—caring about people, kindness, and empathy—are closely related. Figure 9.2 synthesizes the results of dozens of research studies investigating caring, kindness, and empathy individually. Though they are individual concepts, they affect critical employee and organizational issues in similar ways, and Four-Star leaders make it a practice to cultivate all three.

SHOULD YOU THROW THE BOOK AT THEM?

When Barry McCaffrey was a lieutenant colonel commanding a battalion, the Army was fastidiously tracking metrics on any and all activities using the green, amber, red "stoplight" system. One of the things being tracked was books returned late to the library on post. One day, he

★★★☆

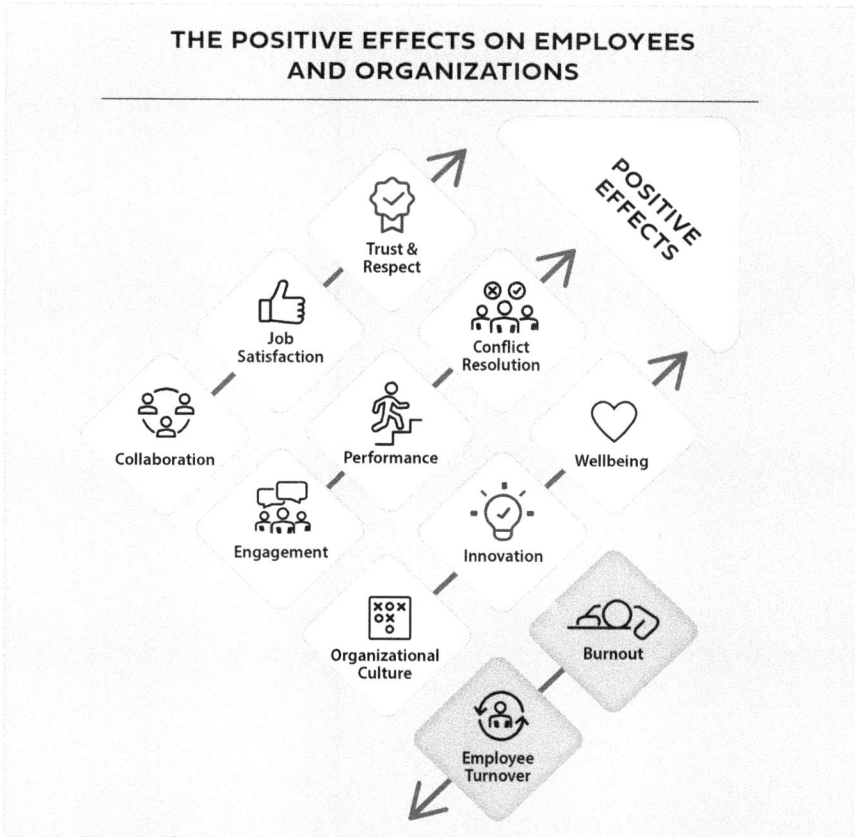

Figure 9.2. The Positive Effects on Employees and Organizations When Leaders Care About Their People, Are Kind, & Show Empathy. All three individual acts (i.e., caring, kindness, & empathy) result in the same effects.

was notified that one of his second lieutenants had checked out a book and had not returned it. General McCaffrey described this particular lieutenant as a big, amiable, very competent soldier with a biochemistry degree from Yale University, and the book in question was a tome about chemical warfare against the Soviets in Afghanistan.

> [The] S3 (operations officer) talked about [it with] him. [The second lieutenant] said, "Oh. Okay, sir, I'll take care of it." Then we got another late book [notice] on the same book, and it came to my attention. I said, "What's with the late book?" [The S3] said, "Oh, I just talked to him. He said, no, he turned it in." Then there was a third one that came in, and I said, "Go ask him what the hell's going on, or ..."

★ ★ ★

The S3 again spoke with the lieutenant who said, "Sir, I went under pressure. I did turn it in. I took it over there, and they were already closed. So, I slid it under the door." Afterward, the S3 called the post librarian to determine what was going on. The librarian explained there was no way the lieutenant could've slid the book under the door because it was far too thick to fit through the gap beneath the door. So, the S3 escorted the lieutenant to McCaffrey's office.

> So, he walks in with the lieutenant. I said, "What's going on? What does this mean?" [The lieutenant] started crying. He started boo-hooing. He lost the book and went into a panic. He was new to the Army, new to Germany, new to the battalion. He went in, and he started crying in front of the two of us. So, we had to comfort him, and we had to explain to him why trust was so important to us in the military. We had to believe each other's words, and could he learn? So, he learned from it. We didn't fire him. We didn't put a scarlet letter in his file.

When then-Lieutenant Colonel McCaffrey could've meted out serious discipline to the second lieutenant, he chose, instead, to be kind, and the young officer learned from it.

In a traditional hierarchical structure governed strongly by regulations and procedures, it could be easy to believe there is no need to be kind. Everyone is expected to operate as instructed, without worrying about individual differences and needs. Automatons are safer. There's less perceived risk of people getting out of line, requiring extra effort, or generating human resource controversy. There's also less perceived personal risk on the part of those in leadership positions because they don't have to invest themselves in relationships with their people. It can all be transactional. Things can be kept distant, objective, and sterile—"It's not personal; it's just business." Numerous traditionally hierarchical industries reflect this: healthcare, academia, manufacturing, and the federal government, as examples. For many, the U.S. military may be the first example of such a hierarchical organization. Yet, half of the Four-Stars expressed the importance of being kind to those they led. Why?

Before further considering the role of being kind in leadership, it is worthwhile to stop and consider a basic question: What does it mean

★ ★ ★

to be kind? Sometimes when we try to define things, we frame it in terms of what is opposite to it. For instance, we might define dark as the absence of light or happy as the absence of sadness. Similarly, with being kind, we might be tempted to say it is the absence of being mean, but that's inaccurate. Being kind is more than the absence of meanness. Being kind is active. It has substance. It involves doing something appealing or charitable for or to another person, without expectation of repayment. That is, being kind isn't transactional. The moment it becomes transactional, it is no longer an act of kindness, rather it is a service rendered. Similarly, if someone is being kind with the expectation of reciprocity, it isn't truly an act of kindness; it is the first step in an exchange. Being kind isn't something you do to receive something in return, it is, as General Lori Robinson suggested, "being a human being. [...] it's treating people the way you want to be treated," because it is the right thing to do.

Despite the Four-Stars' clear appreciation of the value of kindness in leadership, it has been overlooked until relatively recently. Only a couple of books have been written about it, and almost all the limited number of scientific studies on it have been published in the last 20 years. The question is "Why has the importance of kindness to leadership been overlooked?" While the answer is uncertain, there are some things that may have contributed. Perhaps it is due to the strong, lasting influence of Frederick W. Taylor's Scientific Management theory, where people are viewed as equipment to be used as efficiently as possible. Such "equipment" doesn't need acts of kindness. It could be that the authors of popular and scientific leadership literature had other interests and simply never considered the role that kindness could play in effective leadership. Maybe being kind has been overlooked due to rigidity of some human resource management systems that view treating people as individuals as an intolerable act of unfairness. Further, it is likely some leaders are concerned that being kind could be viewed as a sign of weakness. It is possible it hasn't been addressed because people have generally understood the value being kind has in essentially all human relationships, including leadership. Though if that were true, it's hard to reconcile it with the reality of the significant

★ ★ ★

number of unkind people in the world who are in positions of leadership. Whatever the drivers for the delayed recognition of the importance being kind has in leadership, it is increasingly clear that it has tremendous value.

Reshape our brains and our leadership with kindness

While there are people in the world who are perpetually mean, most people recognize that being kind is both valuable for relationships and, simply, the right way to treat someone. We like it when people are kind to us; it feels good. As mentioned earlier, there's a neurohormonal reaction that occurs when people treat us kindly. Just as when someone cares about us, when someone is kind to us, our brain releases oxytocin.

Oxytocin is commonly referred to as "the love hormone." It generates positive feelings, trust, and connection between people. In fact, it is responsible for what may be the strongest of human bonds—that between a mother and her newborn. What's more, oxytocin has direct impacts on neuroplasticity—our brains' ability to remodel, rewire, and recover. That is, it can literally reshape the way the brain is wired and functions, altering how we perceive the world. For instance, in animal models, exposure to oxytocin decreases post-traumatic stress disorder symptoms, depression, and autism-spectrum behaviors. Additionally, in numerous animal and human studies, oxytocin decreased stress and anxiety-related behaviors, among other benefits. Consequently, when someone does something that makes our brains produce oxytocin—like appreciating us, being kind to us, or loving us—we will be more at ease and can perceive the world in a more positive and trusting manner. That mental state promotes the development of strong, meaningful relationships.

The positive neurohormonal effects of being kind to someone extend to relationships among leaders and those we lead. Leadership studies have shown when leaders are kind to those we lead, there are significant improvements in trust, loyalty and retention, job satisfaction, employee engagement, team performance, organizational culture, and well-being and mental health, as well as reductions in

★★★☆

stress and burnout. Overall, these are hugely positive effects for leaders and organizations, and they come at little-to-no cost to either. Given what we know from our own experiences, as well as what has been demonstrated in scientific studies, leaders who are not intentionally engaging in being kind to their people are a) not treating people rightly, b) missing the tremendous organizational benefits of being kind, c) are fundamentally derelict in their duties, and d) probably shouldn't be in any sort of leadership role.

It can be easy to overlook the impact an act of kindness has on someone else. This is particularly true for leaders. Unfortunately, many people do not expect to receive kindness from those in leadership positions, so what may seem like an inconsequential act of kindness on the leader's part can have a lasting positive effect on the person receiving it. This was something General Les Lyles experienced at an event the week preceding his interview. "I can't tell you how many people came up to me and said that a simple word of kindness or giving one of my coins to somebody literally 20 years ago [...] was so significant to an individual for his or her career or [for] just getting by." Because General Lyles sees being kind as part of being a good human and good leader, his kindness hadn't seemed notable to him at the time. However, those acts of kindness stayed with people for decades and cemented him in their minds as a great leader. General Chuck Wald also recalled failing to recognize the impact of what he considered a small act, that in his own eyes was simply the right thing to do.

> I had a kid one day, I don't know where I even saw him, [...] he had been an airman in one of the bases I commanded overseas. Somehow, [...] on one of these dreary nights [...], I picked him up and gave him a ride because he was walking. I had my military vehicle. About 10 years later he wrote me a letter and said how much that meant to him...

You never know how much a seemingly small act of kindness can affect someone, and most of the time it doesn't cost you anything. Reflecting on the current state of society and the value of being kind, especially from people in leadership roles, General Lori Robinson summed it up well: "If in our world, we were more worried about raising everybody

★ ★ ★

up than tamping everybody down and talking crap, it would be a whole different place."

General Robinson recognized perhaps the biggest reason people fail to be kind to others—egocentrism. If we are focused on ourselves and our own greatness, our tendency is to view others as a threat. When we are threatened, our brains revert to the "fight, flight, or freeze" state, with many of us choosing to "fight." As a result, we often "fight off" the threat by being unkind with our words or actions. That isn't what good leaders do; it's what mean children do. Four-Star leaders realize it isn't about them (re-read Chapter 3) and make it a practice to be kind to others. One way they are able to accomplish that is through trying to view things from other people's perspective.

> ## SELF REFLECTION
>
> Can you think of a time when someone in a leadership role was kind to you? If so, how did that make you feel toward them? How much did it cost them to be kind to you?

WHAT DOES IT FEEL LIKE IN THEIR SHOES?

During the first Gulf War, Gene Renuart was a lieutenant colonel flying combat missions in the A-10 Warthog. At the time, the Air Force had decided to put its most experienced pilots with the youngest wingmen in what could be seen as a mentoring relationship. Then-Lieutenant Colonel Renuart was paired with a new lieutenant named Eric, and they flew together for almost every combat mission. Through the preparation, flying, and debriefing of the countless missions over Iraq, they developed a close relationship. General Renuart recounted a harrowing event where that relationship was particularly important.

> Late in the war, a friend of [Eric's] had been flying up in Iraq and got hit by a surface-to-air missile, and the airplane was in bad, bad shape. The

★★★

guy should have jumped out of the airplane. In the A-10, we had a back-up flight control system that could allow you to fly the airplane, not on hydraulics, but rather on cables. It was designed to get you out of harm's way, not to fly 200 miles back to an airfield and land. This young guy decided that he was going to try to bring this airplane back.

As the young pilot discussed it with the operations center monitoring the situation, he received tacit support from some of the supervisors who advised him to proceed if he felt he could make it back to base. As the wounded A-10 was returning, Renuart and his wingman were in their A-10s.

My young wingman and I are holding number one on the deployed airfield, getting ready to take off, to go back on another combat mission. This kid's coming in on final land. The problem is this system, the flight control inputs get more erratic the slower you get; they're less effective. He had lost enough of the flight controls in the part of the aircraft that he didn't have elevator authority to land the airplane properly. So, he came in and basically at 120 (knots) or so, loses control of the airplane, because there's just nothing left and slams down into the overrun. So, [at] about 11 o'clock on our canopies (slightly left of straight ahead), we see him hit the runway. [He] comes up over top of us, inverted, and then hits the ground about 300 or 400 feet down the runway.

Set to take off on their own mission, Lieutenant Colonel Renuart's young wingman had just watched one of his best friends die in a shocking and terrible crash.

So, I looked over at Eric and I said, "Hey, Eric, let's go home." He goes, "No boss, we've got a mission to go fly. We're going to go do that." I said, "Okay, how are you feeling?" "Well, I'm okay." I said, "Okay, here we go. We're going to take off." So, we take off, and we get airborne, and I know he is not doing well. I said, "Okay, just follow me. We're going home." Just to get him back [and] settled down to let him collect his thoughts and all that had happened. I mean, he just watched one of his best friends die. [As a leader, you have to understand] that there are times when you have to say, "Surrender." You have to say the better decision is [...] "This is not the right time for that kid to be out here trying to figure out what he just watched." So, we took the airplane home and put it on the ground, and he flew again a couple days later. He and I went out and flew, kind of our last few combat missions together at that point, and he did well.

★★★

A lot has been written about the role of emotional intelligence in successful leadership. Empathy is one of the principal components of emotional intelligence, but for many people it is often confused for sympathy. While sympathy is having compassion and caring for someone in need, empathy is taking on, to some degree, the emotional and/or psychological state of another person. To truly empathize with someone means being willing and able to move beyond acknowledging their situation to entering into it psychologically and feeling the weight of it. Empathy is the ability to step out of our own perspective and into that of someone else. General Stan McChrystal described it as "being able to put yourself in the other person's shoes to the point that you appreciate what they want, what their perspective is."

> Empathy is understanding that [those you lead] don't see it like you do. You're seeing it from your senior position. [...] They're seeing it from grunt [level], and they're going to have different ideas and equities, and you've got to be able to get it down there and at least understand them.

As with caring about people and demonstrating kindness, there are multiple leadership benefits derived from empathy, including increased employee satisfaction, engagement, well-being, trust and respect, innovation, team collaboration and communication, organizational culture, conflict resolution, and decreased employee turnover. However, one benefit stood out from the others: perspective. Empathy—the ability to put ourselves in the shoes of others—opens us up to seeing things from a fuller perspective than we can on our own. The reality is, we may think we can see things from the perspective of those we lead, but that's not true. Studies have shown that people in leadership positions have a decreased ability to understand and learn from the perspectives of those we lead. Some of the numerous contributors to this inability include power distance, cognitive biases, organizational structure, and a lack of emotional intelligence, among other things. There are negative consequences that derive from our inability to see things from others' perspective. General Stan McChrystal put it simply: "If you're trying to get—whether it's young American soldiers or Afghan tribal leaders or anybody—to do something, and you can't

★ ★ ★

appreciate what they see and why they see [...] things that way, then you're dead in the water."

Leaders must have the broadest possible perspective; this perspective allows you to see the potential threats and opportunities for the organization. It allows you to have a vision for the future, while also seeing where the organization has come from. But having an expansive perspective can be hard because, as individuals, we are limited to our own point of view unless we can tap into the perspectives of others. Beyond being caring and kind, this is where empathy can be exceptionally beneficial. If you are willing and able to empathize with those you lead, your perspective will be broadened to include theirs, whether it's the threats that concern them or the opportunities that excite them.

In addition to broadening a leader's perspective, empathy has several positive effects on both those we lead and our own leadership. (Figure 9.3) It shows others you care about them because you understand and share their feelings.

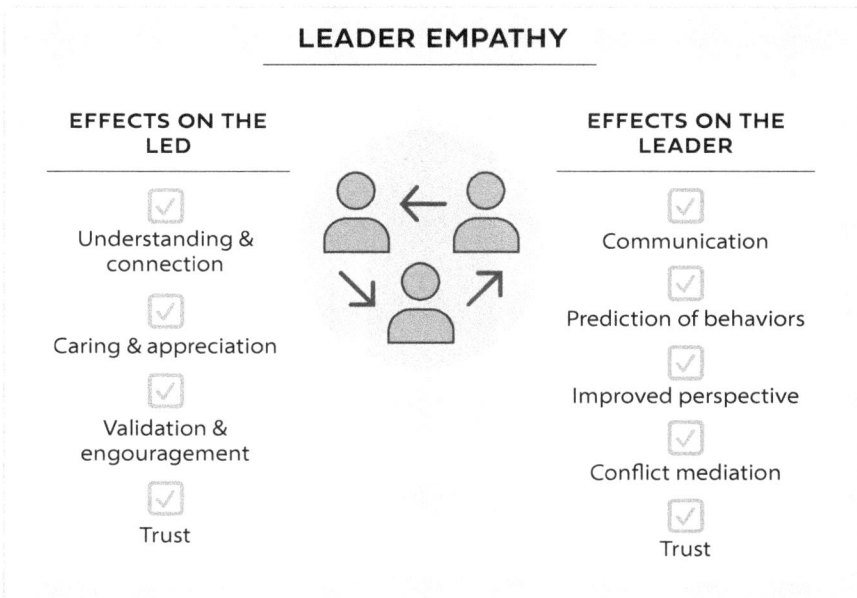

LEADER EMPATHY

EFFECTS ON THE LED	EFFECTS ON THE LEADER
Understanding & connection	Communication
Caring & appreciation	Prediction of behaviors
Validation & engouragement	Improved perspective
Trust	Conflict mediation
	Trust

Figure 9.3. The Effects of a Leader's Empathy. When a leader empathizes with those they lead, there are positive effects for both the led and the leader.

★ ★ ★

As shown in the figure, when we, as leaders are empathetic, it builds understanding and connection with those we lead; demonstrates caring and appreciation to them; and validates and encourages them. Those effects, then, build trust and foster strong relationships, which are all fundamental to the ability to lead others. However, empathy also has effects on us as leaders. It improves our communication potential, because empathy gives us a greater understanding of where our people are psychologically, allowing us to frame our communication more effectively. Further, when we understand others' emotional and psychological state and perspective, we are better able to predict their behaviors and respond appropriately. Additionally, when we empathize with the concerns of others, we can mediate conflicts more effectively and find solutions for underlying issues. Communicating effectively, predicting behavior, and effectively mediating conflicts are all extraordinarily valuable leadership capabilities. These can be particularly helpful for avoiding misunderstandings and problems that might otherwise derail effective team and/or organizational functioning.

★ ★ ★ ☆

COMMUNICATION

THE FOURTH STAR OF LEADERSHIP

CHAPTER 10

COMMUNICATION:
THE MEDIUM OF LEADERSHIP

You've got to communicate. You can have the best ideas, the best plans, the best leadership philosophy, but if you don't communicate that to your people on a routine basis, they're not going to understand it. They won't be able to support it, and so, I think it's absolutely essential.

– JAMES CONWAY, U.S. MARINE CORPS –

In 1969, then-Captain Johnnie Wilson was a young company commander in Army logistics serving in Vietnam. The logisticians played a crucial role to Army operations because they were solely responsible for supplying soldiers with the food, weapons, ammunition, and equipment they needed to fight the war. However, there were times when that did not happen.

> We had some very important missions being logisticians: to get supplies to where they should be [...] But, [...] we had a couple occasions where we were going to deliver to point [A], and really the requirement was in [B] or [C's] location. During that time, things were moving so fast, but [...] if we had taken the time or knew that communications was really the issue, [it would have been different]. We should have confirmed [...] the location. So, I [asked myself], "Wilson, what did you do to help the situation or to help the success of the operation?" Every other day, we should have sat and said, "Okay, B Company, 123rd is going to be here, and somebody, a lieutenant, needs to check with the support leader of that element to

★★★★

make sure [they are going to be there], and if they're on the move, we need to know about it."

But, the logistics personnel did not make it a consistent practice to communicate with the soldiers in the field to determine where they would be when the supplies would be delivered. Consequently, those supplies were sometimes not delivered to the right soldiers when they needed them. This problem was something General Wilson saw re-played at other times.

> As the AMC (Army Materiel Command) Commander, the combat guys, my counterparts, would always say, "Wilson, how come you guys can't deliver like Walmart and Amazon?" [...] Well, Walmart and Amazon, they have a location. They don't have to worry about troops moving at midnight or two o'clock in the morning, where[as] the Army moves.

Communication—the final crucial element for Four-Star leadership—is, on one hand, the most often overlooked contributor to leadership success and, on the other, arguably the most important factor in suc-cessful leadership. Without effective communication, things will not get done, or, as in General Wilson's example, they will get done poorly with high risks to the organization. His example highlights that com-munication must be a multifaceted, two-way process if it is to be suc-cessful. The logistics personnel didn't ask where the companies in the field would be; they assumed they would remain where they were. On the other side of the failed communication, those units in the field did not inform the logisticians where they would be. Neither side was seeking, sharing, or listening for information until Captain Wilson identified the problem and instituted a solution. The complexities and pitfalls of the communication process will be covered in more detail in Chapter 12. For now, we will cover the topic of communication more broadly.

Leaders must be able to communicate if we have any hope of in-fluencing those we lead and, in turn, accomplish the mission. No Four-Star was more adamant about the centrality of communication in effective leadership than General Les Lyles, former Commanding General of Air Force Materiel Command. When I asked him what the

★★★★

three most important keys for leadership success were, he answered, "Communicate, communicate, communicate." General Lyles later expanded on this comment.

> [...] in times where I've gone back and said good communication was the key to [...] success, it wasn't like upfront. [...] communication is just [...] part of the culture that you build [...] and that transparency and openness with people [...] turns out, to me, to be the best way [...] to address and solve things.

As important as the other Cs of leadership (i.e., Character, Competence, and Caring) are to leadership success, it could be argued that if you cannot communicate with those you hope to lead, you won't be able to lead them. No matter how great your character, if you cannot communicate that to another person, whether via verbal or nonverbal means, it will have no influence on that other person. The same is true for competence and caring. If you cannot communicate your competence or that you care, those things will have no influence on those you wish to lead. Conversely, we need not look far at all to find people in leadership positions who wield tremendous influence in the absence of legitimate evidence of character, competence, or caring about those they are leading. Consider certain U.S. politicians as examples. While, perhaps, the antithesis of Four-Star leadership, those people exemplify the power of communication. Without character, competence, or caring, they have obtained influence and leadership positions through their ability to communicate and resonate with significant swaths of the population and incite those citizens to follow them. But their influence is always short-lived. While they may be able to communicate effectively to stir people to action, that action will always be to ends that demonstrate their lack of character, competence, and caring. When their lack of these crucial qualities is revealed, people will stop following them.

Communication, then, is fundamental to leadership, and it carries multiple benefits. Leaders who can communicate a vision, provide guidance, and listen to team members can cultivate the trust essential to teamwork and improve organizational effectiveness. Additionally,

★ ★ ★ ★

skillful communication fosters understanding, alignment, and motivation amongst the group. Through their ability to communicate well, Four-Star leaders unify, energize, and sustain those they lead to accomplish the mission. The ability to communicate well was something General Gene Renuart discussed.

> Being an effective, humble communicator [...] allows you to establish a relationship with your peers, your superiors, and your subordinates in a way that they can respect. Then when you transmit, you want to be as clear and as concise, yet specific as you can so that the message gets across without lots of confused looks on people's faces. That's important whether you're talking to a small team or you're sending out an operations order that's going to send 150,000 people in a direction. [...] that ability to communicate, I think, is really, really important.

The crucial role and numerous benefits of effective communication make it, arguably, the bedrock of leadership.

SELF REFLECTION

Do you put as much effort into communicating clearly with those you lead as you do with making plans and decisions?
Can you think of a time when there was an organizational failure due to poor communication?

WHAT ARE THE BUILDING BLOCKS OF EFFECTIVE COMMUNICATION?

When we think about leaders who communicate well, it's easy to think only about spoken communication. Our archetype of a leader as a great communicator is usually derived from those who are great orators, the quintessential example being Dr. Martin Luther King, Jr. However, spoken communication represents only one medium, and there

★★★★

is much more to effective communication than simply the words that are spoken.

All communication involves at least two people, a person sending a message and a person receiving a message. However, for this discussion, the focus is on what we can control, which is limited to ourselves. That is, while we can consider things about the other person that will influence how they might receive our message, we can only control our side of the communication. While perhaps not exhaustive, I have identified at least 12 communication principles over which we can have some control. (Figure 10.1) Those principles fall into two categories: communicator traits and communicator actions.

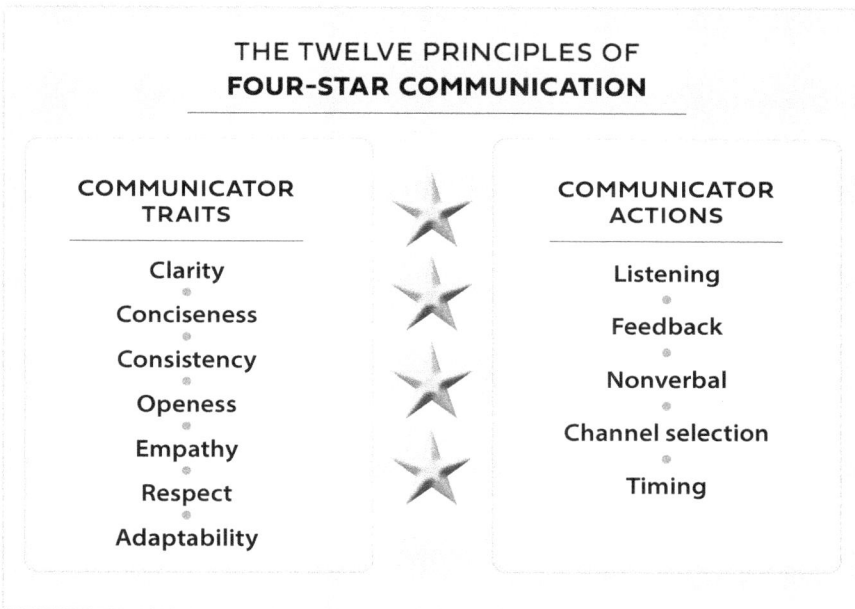

THE TWELVE PRINCIPLES OF
FOUR-STAR COMMUNICATION

COMMUNICATOR TRAITS	COMMUNICATOR ACTIONS
Clarity	Listening
Conciseness	Feedback
Consistency	Nonverbal
Openness	Channel selection
Empathy	Timing
Respect	
Adaptability	

Figure 10.1. The Twelve Principles of Four-Star Communication. These are all factors in communication that are within one's control. For instance, we can be clear and consistent, adapting to the person, environment, etc. We also can listen intently, time the communication appropriately, and make sure our nonverbal communication fits with what we are saying.

Instead of thinking about the principles as traits to acquire or things to start doing, it is better to think of them as traits we all have and things we all do—we may just need to improve upon them. If so, that means

★ ★ ★ ★

everyone can improve their communication capabilities. This is good news, because the more of the principles we can improve and master, the higher the likelihood we will be able to communicate well. In going briefly through the principles of communication, instead of thinking as an orator, it's more likely to be helpful to think of yourself speaking with a friend or colleague about something important.

Four-Star Communicator traits

As shown in Figure 10.1, there are seven communication principles that fall into the category of communicator traits. These are not physical attributes, but rather represent a person's mindset toward communication.

Clarity is essential for effective communication—if the message you intend to communicate is not clear, then you have failed to communicate it well. To communicate well, the message must be clear and easily understood. The best communicators—Four-Star communicators—do not try to impress their audience with long and complex or obscure words. Instead, they use straightforward language free of jargon or ambiguous terms. General J.D. Thurman provided a great example of this. He emphasized the need for clarity in communication to the general officers he developed.

> Keep plans and instructions or orders simple. Simple. [...] If you're trying to give direction, don't put something in there that's got to be interpreted. [...] I can't tell you how many meetings I've been in, and then after a meeting, you have a meeting to decide what was talked about, because nobody understood.

In addition to the words used, the other component of clarity in communication is the logical organization of the message. The words being used may be easily understood, but if they are not arranged clearly, the message can be lost or misinterpreted. There are numerous memes surrounding correct punctuation that demonstrate this. For instance, there is a tremendous difference between "Let's eat, Grandma," and "Let's eat Grandma." To communicate with clarity, make sure the words you use don't require interpretation, and make certain that what you think you are saying is what you are actually saying.

★★★★

Conciseness is a hallmark of skillful communication. A concise message is to the point and free of unnecessary details that may obscure it. That makes a message easy to understand and remember. As General Gus Perna trained others how to communicate orders well, he told them that the best leaders "give orders that are simple, easily said, easily remembered, and easily repeated." That is excellent guidance for making sure your message is concise.

Consistency of the message is also essential for effective communication. People who are inconsistent in their communication may say one thing and do another, or they may say one thing at one point and then say something completely different later. There are a couple of ways such inconsistency creates dissonance in our signal, disrupting communication and our ability to lead others. First, when we are inconsistent with what we are saying—when our messages don't line up—people don't know what the correct message is, what they should believe, or what they should be doing. Mixed or contradictory messages destroy meaningful communication. Perhaps more importantly, when our message is inconsistent, whether through the words used or the misalignment of our words and actions, people will not trust us. That dissolves our ability to lead. General J.D. Thurman again emphasized the need for consistency.

> [I would tell them], you've got to keep things simple and be consistent in your messaging, and you've got to constantly revisit that. So, people know, "This guy knows what he's talking about. I believe in him. He's not a wishy-washy guy, and he's a guy that we can go to and he's going to give us direction."

Openness is being open to or receptive toward other ideas and experiences—being open-minded. Some readers may recognize it as one of the "Big Five" personality traits, and it has been associated with increased leadership effectiveness. Openness in communication is a two-way street. In one direction, it is the willingness to share information with others. The other direction is the willingness to consider, and even be persuaded by, others' perspectives. This is the principal subject of the next chapter. For now, it's worthwhile recognizing our

★ ★ ★ ★

ego and the value we place on our own ideas can bias us and make us closed-minded to others' perspectives. When we can overcome those biases and maintain an attitude of openness, we have a trusting attitude of others and are approachable. Consequently, our people will be more likely to share vital, though sometimes negative, information we need to be successful.

General Frank McKenzie, *on Openness and Approachability*

From March of 2019 until April 2022, General Frank McKenzie's Marine Corps career culminated in his service as the 14th Commander of U.S. Central Command. Throughout his 43 years of military service, General McKenzie led in myriad roles where the full communication of vital information—both bad and good—was critical to mission success, and approachability was instrumental in achieving it.

> You've got to be willing to be open to the flow of information. I found that, in history, a lot of senior leaders become unapproachable. They don't like to get bad information, or they shoot the messenger in the face. So, here's the thing, if you're a Four-Star or a Three-Star or any commander and you're in the habit of jumping down the throat of people who give you bad news, people still have to give you that news because they're obligated to do that. But they will shade it in a way to try to minimize getting hit in the face, because nobody likes to do that. So, what you need [...] as a leader, is [...] the unvarnished truth. [To get] that, you've got to be honest; you've got to be approachable.

Being approachable and open to receiving bad news did not necessarily come easily to General McKenzie. He admitted having a natural tendency toward a significant temper, and he worked hard throughout his career to become welcoming of negative information. One method he used to do so was to inform his personnel that they should not be afraid of surprising him with bad news.

> [As the Commander of Central Command] I [would] tell people, "Look, I was the Military Secretary to the Commandant of the Marine Corps, two Commandants of the Marine Corps. I was the Director of the Joint Staff. There's no bad news you can give me that I haven't heard before. I mean, [there's] nothing a human being can do to another human being that you can tell me that I haven't heard before. Nothing. [There's] no operational

★ ★ ★ ★

event you can tell me that I haven't got somewhere in my background [and have] heard about it."

By addressing it at the outset and allaying his people's fear of how he would react to bad news, General McKenzie fostered open and direct communication of all information, whether good or bad. Doing so helped him to be a much more effective leader.

Empathy is an instrument, perhaps even a neuroscience-based superpower, for becoming a Four-Star communicator. As covered in the last chapter, empathy is the ability to consider and understand the feelings, perspectives, and needs of others. When we understand the cognitive and emotional state of those we are communicating with, we can "speak to them"—truly communicate—on a much more powerful level. When we are empathetic with others, our brainwaves actually begin to align with theirs—in those moments, studies have shown that we think similarly, even to the point of actually feeling their physical pain. This all contributes to us feeling closer to those people we empathize with. As a result of these effects of empathy, we share their perspective, which allows us to shape how we communicate. Consequently, we can communicate more effectively while fostering understanding, building connections, and preventing and resolving conflicts.

Respect is an easily overlooked component of effective communication. It underpins our ability to empathize with others, to be open to and listen to what they have to say. When we don't respect the person we're trying to communicate with, they recognize that. We may think we can disguise it, but our disrespect comes across in our nonverbal communication. When people feel disrespected, they can become resentful and close off toward us. That leaves us with little to no hope of having effective communication.

General Frank Grass saw lack of respect as an issue at the root of several societal problems. He surmised that many people fail to recognize that there is a difference between showing respect to someone and agreeing with them on a given topic. As a result of the loss of that

★ ★ ★ ★

recognition, effective dialogue is disrupted, and our society suffers negative consequences.

> I look across the nation, and people are attacking other people for [having] a different view. [...] respect doesn't mean you have to agree with someone, but you respect their opinion. Don't go after them as a person. Have a debate and a discussion about the opinion, because you're going to have your own opinion and there'll be a disagreement. [...] It's not any different than [with] an enemy. We respect our enemy. When we grow up in the military, we study how the enemy thinks and how they fight. We respect that. We don't agree with it, but if we ever have to go into battle... It's all about respect in so many different terms, and I think that's one thing today that we've lost.

Adaptability in communication is the ability to tailor the message and medium to the needs and preferences of the audience. Four-Star communicators can shift their communication style to suit different individuals and situations, and they do so in the moment. Admiral Jim Loy prepared specifically to be able to adapt his communication in real-time.

> I may have prepared a 20-word answer to that question I was presuming was going to get asked, but if it was asked in a slightly different way, if it went in a slightly different direction, I needed to have the command of the subject matter to not only provide the answer I wanted him to get, but I also wanted to be able to provide the answer that he wanted to hear [...] to fulfill the search for information that he was looking for. So, [preparations for] both [of those possibilities], I think are very appropriate. Prepare [...] well enough to adapt to the line of thought being presented in the conversation.

Four-Star Communicator actions

Five of the communication principles represent actions that facilitate effective communication. They require preparation and active engagement on the part of the person communicating.

Listening may be the most difficult part of real communication. It requires the self-control to stop talking, pay attention to what is being said, and refrain from forming our rebuttal. Yet, as difficult as it is, it is central to communication; the next chapter is dedicated to it. If we

★ ★ ★ ★

cannot listen to and understand the other person, we won't be able to truly communicate—instead we will only be broadcasting.

As just previously covered, Four-Star communicators understand and empathize with their audience, which demands careful, attentive listening. They know communication is a two-way process that requires active listening. By doing so, Four-Star leaders demonstrate that they value the input of others, nurturing understanding and trust in the process. Contrast this with those people in leadership positions who think communicating means they should just be broadcasting ideas and instructions to others. Unilateral transmittance like that is not really communication, and it undermines leadership effectiveness and the potential for accomplishing the mission. That is because leaders who refuse to listen fail to have the up-to-date, ground-level information essential for success in large organizations and rapidly changing environments. Would-be leaders who don't listen actively undermine themselves because they never listen long enough to get the vital information they need to make the best decisions. General Gene Renuart sees listening as an especially important part of communication.

> You have to be a very good communicator and probably focus more on the listen than on the transmit. You have to be able to understand what it is people are saying, not so much to you, but rather what the environment is telling you about the challenge that you may have, no matter what it is.

Feedback is a highly valuable component of communication, and, as with other principles of communication, it is a two-way street—receiving and giving. Four-Star leaders value receiving feedback and understand the importance of giving constructive feedback. Receiving feedback well—particularly negative feedback—can be hard and is something that often requires intentional work, as General McKenzie demonstrated. However, getting that feedback is absolutely essential in maintaining an understanding of your environment. General Barry McCaffrey mandated that his people had to give him feedback.

> You've got to make sure your personal staff—your primary reports—understand they're to disagree with you and to provide negative feedback. Then, you've got to listen to what they're saying. You don't have to do it, but you damn sure better listen.

★ ★ ★ ★

Most people in leadership positions are more accustomed to giving feedback to those they lead than receiving input from them. But leaders must be open to listening to what those they lead have to say, and then we need to acknowledge it in a meaningful way. Within the context of communication, that acknowledgement—or feedback on the feedback—carries three benefits. First, it demonstrates to the person who sent the message that we are attentively listening, with the attendant benefits that listening brings. Next, the feedback demonstrates that the message has been understood, and it allows for adjustments if needed. Finally, it allows us to encourage and provide constructive modifications, if needed, to maintain alignment with the overall goals.

Nonverbal communication is a highly important component of communication. Most people recognize, and various studies have confirmed, that nonverbal aspects play a tremendous role in overall communication. Perhaps more than anywhere else, nonverbal communication has the potential to negatively impact our communication clarity and consistency. When we say one thing, but our facial expressions, gestures, and tone of voice say something different, those on the receiving end get mixed signals and will distrust us. We have to be conscientious of what we may be communicating through our body language, expressions, gestures, and other nonverbal signals. These will either complement or contradict the desired message. Four-Star leaders recognize, as General David Rodriguez said, "The audio's got to match the video," and make certain their nonverbal communication aligns with their intended message. Getting that alignment can be challenging. There are a lot of nonverbal signals that we are completely unaware we are sending. Consequently, an outside observer is often required to identify and help improve the alignment between verbal and nonverbal communication.

Selecting the right channel for communication is increasingly important, and there are steadily increasing modalities to choose from. Whether the communication is face-to-face (e.g., spoken or sign language), written (e.g., prose, poetry, texts, or posts), digital (e.g., audio or video), or symbolic (e.g., artwork, photographs, memes, etc), Four-Star

★★★★

communicators adapt the channels they use to communicate. By understanding the nature of both the audience and the message, great communicators are able to use the best channel to ensure effective communication and get their message across.

Appropriate timing is the final of the twelve communication principles, and it can often be neglected. Delivering the right message at the wrong time means it will be unlikely to be effective or well-received. In fact, poor timing can have tremendously negative impacts on our effectiveness as a communicator, because poor timing can make us appear callous or clueless, either of which are not conducive to successful leadership. Some circumstances, such as during a crisis, demand immediate communication. Others require careful timing of delivery to optimize the message's impact.

By being conscientious of the twelve principles of communication and actively cultivating them in our own communication efforts, we can greatly increase our leadership capability and effectiveness. The next chapter will focus specifically on the most difficult part of communication—listening—and the final chapter in this section will bring together all of the components of communication to help you understand that it isn't what you say that is important, but rather what they hear that matters.

★ ★ ★ ★

CHAPTER 11

LISTEN LIKE YOU WANT TO CHANGE YOUR MIND

Listen with a willingness to be persuaded. [...] Listen so well, you hear also what's not being said. [...] by listening to them, you understand where they think they can contribute, what they can do, where their worries are.

– GENERAL JIM MATTIS, U.S. MARINE CORPS –

When General Barry McCaffrey was a Major, he was stationed in Germany as a brand-new battalion commander in the 3rd Infantry Division. The sergeant major who worked with him was a World War II veteran who impressed on young Major McCaffrey the need to listen to his soldiers.

> He came, and he sat down, and he said, "Sir, I'm going to tell you something [you need to hear] as a battalion commander." He said, "When somebody calls you and says, 'Major, I just drove by your motor pool, and it's all on fire. There's a giant fire going on out there.'" He said, "Your tendency's going to be to say, 'No, it isn't. It isn't on fire, and if it is on fire, I didn't do it. And if I did do it, then I've already got a solution ongoing.'" He said, "What you've got to do is say, 'Sergeant Major, thank you for bringing that to my attention. I'm going to go look into it.'"

The sergeant major's wise counsel shaped how General McCaffrey responded throughout the rest of his career. He realized he had to not

★★★★

only listen, but also seek out and solicit the perspectives of his soldiers, especially those that might differ from his. When Four-Star leaders listen to others, it isn't just to get a correct perspective or answer. In some cases, listening allows them to understand something behind the feedback. General McCaffrey concluded his comment, saying, "When somebody provides negative feedback, always listen. It may be wrong, but there's some reason they're telling me that."

Many people in leadership positions fall prey to a common fallacy— that we are supposed to know everything and have all the answers. That outlook is an Achilles heel for many who could otherwise be highly effective. Instead of expending large amounts of energy trying to maintain a façade of infallibility, *trying to look right*, they could be open to the inputs of others and *actually be right*. Four-Star leaders aren't worried about being the one who comes up with the right answer; they only want the right answer to be found, no matter who gets the credit. Those leaders know that no matter how skilled and experienced they may be, it is mathematically impossible for them to make the best decision 100% of the time. The only possible way to get the best solution is to listen to and be open to the ideas of others. Four-Star leaders are much closer to being collectors of the insights and ideas of others than they are to being lone geniuses generating answers in isolation.

There is much made of the need for leaders to listen, perhaps because many of us don't. Listening—carefully considering, weighing, and being willing to be persuaded by another person—demands humility. This may be why so many in leadership positions can't seem to bring themselves to listen. First, listening to someone else's ideas requires the willingness to admit you don't have all the answers. That can be a threat to some people's egos and sense of self. It can create deep cognitive dissonance, because listening to others' ideas when they are different from ours forces us to confront the flawed image of our own capabilities—"If someone else is right, and I am wrong, that means I'm not as smart, capable, effective, etc. as I believe I am." That's a tremendously difficult thing to do. When faced with thoughts like that, our brains often minimize the psychological discomfort by attacking the ideas and logic of the other person—we become defensive of our ideas

★ ★ ★ ★

and perspective, don't listen, and ensure the other person will not bring other ideas to us in the future. This problem befell Henry Ford. When numerous people tried to convince him to expand the color options for the Model T, he famously said, "Any customer can have a car painted any color that he wants, so long as it's black." It wasn't until Ford had lost significant market share to General Motors, which offered customers numerous options including choices of paint colors, that Ford began offering the Model T in four colors, including black. By that point, General Motors had already leaped ahead of Ford in sales, a position it maintained for the rest of the century.

Another ego-related driver hindering many leaders is the fear that if we don't have the answer, we will erode people's confidence in our leadership. In reality, only delusional leaders think they know all the answers—the people they are leading know well that the leader does not know everything. So, we need not worry that listening to others will compromise our veneer of infallibility; it won't. To the contrary, a large study of over 4,000 leaders conducted by Zenger Folkman in 2022 showed that a leader's effectiveness with building trust and fostering relationships is directly related to their listening effectiveness—good listeners are better leaders. Thus, contrary to popular belief, listening to those we lead is not detrimental to our perceived capability and authority. Instead, it actually helps us. Four-Star leaders know the power of listening and seek to harness it to improve their leadership efficacy.

WHAT ARE WE LISTENING FOR?

As the first female Four-Star Combatant Commander in U.S. History, General Lori Robinson simultaneously commanded U.S. Northern Command and North American Aerospace Defense Command (NORAD). Prior to that role, when she was Commanding General of Pacific Air Forces, she was responsible for defending 52% of the Earth's airspace. However, instead of wielding the power of her positions, General Robinson found that listening was one of the most important things she could do. "The power of listening is incredible," she said. "People expect that you're going to go in there and change things right

★ ★ ★ ★

away, and I never did because I didn't know what I was getting into. So, it was important to me to watch, learn, and listen."

General Robinson found that listening not only helped her learn, understand, and lead better, but it also demonstrated to people that she valued them, their time, and their efforts.

> I've also had people say, "Ma'am, they're not ready for you, [for some reason or other]," and I would go, "But it's on the schedule." "Yeah, but we'll cancel it." And I'd go, "Do you know how hard they worked to get it in front of me? Let's just have it. I'm not going to yell at anybody..." So, that power of listening, the way it makes people feel, [...] you learn so much more by listening than you do by talking.

Listening carries a lot of leadership benefits, and Four-Star leaders capitalize on them. Those you lead understand that listening is vital in complex situations, when different perspectives and experiences are needed to identify effective solutions. When you listen, it assures people your goal is to get it right, not to be right—and that intention increases people's trust in you as their leader. By listening, you can take in data and perspectives from numerous people who, in many cases, are much closer to the problems and their attendant solutions than you are. In this way, it is possible to learn much more in much less time than would otherwise be possible. Sometimes, though, it may take an investment of time to get to those great ideas. In this regard, Admiral Scott Swift valued not only listening to those who are closest to the problems, but also being willing to continue to listen to them even when their ideas have not previously been helpful.

> [As a leader, you have to be good at] criticizing ideas without criticizing individuals, because that knucklehead, that nine times out of ten has something stupid to say and slows the process down, that tenth time they're going to have that golden nugget. But if you've shut them out of the conversation, you're not going to have access to that golden nugget of knowledge. Now, there's a balance in there, you want to be a high performing organization. But I think creating that mutual respect, that sense of humility, that you're not the smartest one in the room, [is important].

★ ★ ★ ★

Ultimately, listening helps us discover the information we need to make decisions, influence others, and take effective action—in short, to lead. Figure 11.1 presents some of the numerous benefits that occur when a leader listens.

THE IMPACTS WHEN LEADERS LISTEN

Respect

Relationship

Trust

Well-being

Curiosity

Performance

Knowledge

Figure 11.1. The Impacts When Leaders Listen. When leaders listen to those we lead, it shows respect, improves trust, and builds relationships. Curiosity, knowledge, and performance increase, as does well-being. Note that while auditory listening is presented for simplicity in the figure, how someone "listens" may vary by audience. People may "listen" via sign, text, or tactility (i.e., Braille).

Don't listen to respond

As the Commanding General of the U.S. Army Materiel Command, General Gus Perna was dedicated to developing the people he led, including subordinate generals. One of the things General Perna emphasized

★★★★

for his personnel was the importance of listening, and he allowed them to showcase their abilities—especially listening—during high-stakes press briefings.

> One of the things I used to tell my generals was, "Listen to understand, not to respond. Don't put your finger on that mic until you're confident you understand what the question was. Then, I want you to be brief and bright [in providing] the answer. Saying it five times doesn't make it right."

If any of his generals failed to listen to understand, General Perna was quick to remove them from further opportunities in the limelight until they corrected their oversight. "I was draconian about it," he explained. "If they [messed] up those two things, I literally would say, 'Okay, you're not touching the mic for a while [...] You're going to prove to me that you understand what I'm talking about.'"

Apart from inattention or lack of interest while listening, the compulsion to respond is, perhaps, our greatest impediment to listening. Several contributors drive us to respond. The least insidious is the general belief that communication is an exchange of ideas between two or more people. If we accept that notion as true, then we assume there is a need to contribute equally to the exchange. Consequently, it is easy to operate with a reciprocal mindset, expecting to respond to any given statement or idea. Because formulating a relevant response takes thought, the expectation to respond can easily distract us from giving full attention to what the other person is saying. Then, once we have formulated our response, it can feel like a mental hot potato, and we look for the earliest instant to toss it out there, often leading to blurting out a response in the middle of the other person's thought or statement. By doing that, we disrespect the other person and, simultaneously, induce them to do the same thing—listen to respond. As a result, what began as a potentially meaningful conversation rapidly becomes a volley of loosely associated comments or stories from which neither party leaves with new ideas or answers to important questions.

Ego preservation can be another driver of listening to respond. When being thoughtful, well-informed, and intelligent (all traits of most great leaders) are part of one's self-image, that self-image can

★ ★ ★ ★

be threatened when someone presents new ideas or information that either we don't know or that challenges what we believe. In such situations, we have two choices: embrace learning something new and valuable at the expense of being revealed as not knowing everything and having all the answers, or respond with tangential information or, worse, questions aimed at discrediting the other person. As discussed previously, what seems to be lost on many people in leadership positions is everyone else knows they don't have all the answers. So, paradoxically, the more those leaders try to maintain the charade and their egos, the more their deficits are revealed. What they fail to understand is brilliant people aren't the ones who always have the answers, rather they are the ones always asking great questions. Curiosity and a thirst to learn from others characterize Four-Star leaders.

The final reason someone might listen to respond is less negative but can have the same result as ego-preservation. The principal reason most people listen to respond is probably because they are trying to establish commonality with the other person and build the relationship. For example, when Samira is beaming about her new car, Patrick may think he is showing interest and building connection with Samira by telling her about his new car, or conversely, telling her everything he knows about her new car. Unfortunately for Patrick, his sharing can be taken as one-upmanship and, instead of fostering connection it can significantly disrupt their relationship. Patrick has failed to understand the basic situation: Samira was excited about her new car and wanted to tell someone about it. The best thing Patrick could've done was to listen to Samira talk about her car and ask her questions to allow her to talk even more about it. Four-Star leaders understand this and don't listen to respond. Instead, they listen to understand, to learn more about the subject and the person talking. As a result, they come away better informed about both and, in all likelihood, with a closer relationship as well. All of these outcomes are wins for any leader.

★★★★

Questions are the answer

Listening to understand doesn't mean that we just sit quietly and absorb the information someone is delivering. If we are going to fully understand something, especially information or a perspective that is different from ours, we often need to ask questions. But our questions can't be just any questions; they can't be loaded, tangential, or outright irrelevant. When we are focused on listening to understand, our questions must be the right questions, aimed specifically at clarifying the information being received.

During his 39 years of service as a Marine, General Glenn Walters learned how critically important it is for leaders to understand the situation surrounding their decisions. Since no leader has all the information or all the answers, his approach to increasing his understanding was to seek information from others, which he did through asking questions and listening to the answers. When he retired from the Marine Corps, General Walters took this approach into his new role as the President of the Citadel. He recalled how his commitment to asking the right questions and seeking answers altered what happened to a young cadet.

> We had an honor violation about two years ago, when we had some kids take [a professor's] online classes. [The] professor put [a student] up for an honor violation. [...] This professor said, "He said he tried to get in my class, and I never saw his attempt. He's lying." So, she wrote him up for an honor violation [and] gave [it] over to the Honor Court [comprised of cadets]. They're going to do some due diligence, and their due diligence is to call the IT guy from Canvas. [They ask the IT person,] "How do you know he didn't try and get in?" He said, "Well, if someone tried to get in, it would show up on my log that they attempted to get into Canvas and didn't." So that's all, that's as deep as they went. Then they said, "Okay, he's in violation, let's kick him out."

The members of the Honor Court then met with General Walters in his office to discuss the expulsion of their fellow cadet. Before agreeing with their assessment, General Walters had a question for them.

> "Where was he when he was trying to get in?" [The Honor Court members responded,] "Well, he was in his barracks room, sir." So, [I asked], "How

★ ★ ★ ★

many firewalls between the barracks and Canvas?" "What do you mean, sir?" I said, "Well, it doesn't go direct from the barracks to the Canvas firewall; it goes through a router." I knew we had problems with the damn routers in the barracks, and then when you came in the servers in this building [...], it goes from the router there, to the router there, and [then] it goes into the Canvas, and it's successful. So, if it gets past that last firewall, then it tries to get into Canvas, then you'll notice him. But if he never got past those two nodes, it doesn't tell him, "Well, you got stopped at firewall number two."

By asking the right question, General Walters uncovered that there was more to the situation than people had realized. This allowed him to make the correct decision and avoid expelling a cadet who did not deserve it.

Instead of being concerned his status could be threatened by asking questions, Four-Star leaders like General Walters approach others inclusively with a simple, yet powerful attitude based on asking the right questions. Rather than acting as an all-knowing judge, he asked simple, powerful questions, and then he listened to the answers. This gave him the information he needed to reach the right conclusion. Four-Star leaders do this consistently, and it supercharges their leadership by increasing their perspective, understanding, and ability to make the best, informed decisions.

POWERFUL LISTENING ATTITUDE

This person may have the answer to the problem I'm facing. What can I ask them that will help them share it with me?

General Paul Kern, *on Assessing Understanding of the Mission*

General Paul Kern served 38 years in the Army, finishing as the Commanding General of Army Materiel Command. Throughout his career, General Kern used after-action reviews to ask the right questions. He saw those reviews as vital for assessing the successes and failures of

★★★★

different aspects of missions, from the squad level all the way to the corps level. When missions failed, General Kern often found the issue was a lack of understanding of what the mission really was.

> One of the key pieces of [the after-action review] is to ask the people, did they understand what the mission was? So, we spend a lot of time on writing out a mission and an intent statement—"This is what we want you to do,"—and then get interrogated about it by your own staff and then by their own people. These after-action reviews were always [...] focused on that. Quite often you found that the soldier didn't understand exactly what it was he was supposed to do. "Gee, I thought I was just supposed to get to this hill and stay there."

General Kern and his subordinate officers would then use the opportunities to ask further questions with the goal of developing their soldiers' understanding of current and future missions. He would ask his soldiers, "What would you do if the hill was already occupied? What would you do if there was an obstacle in the way of getting to the hill? What are the things that you really need to take into [account during] that planning cycle?"

This process of asking about the soldiers' understanding of the mission and then building on that understanding by asking the right questions was a key focus of the after-action review. The goal was to help everyone understand precisely what the mission was. Those reviews are a longstanding and trusted system for organizational learning, and they are based on the leader's ability to listen.

WHAT ARE THE KEYS TO LISTENING WELL?

As I've previously touched on, it can be hard to listen. A significant contributor to this difficulty is our tendency to prefer to listen to what we have to say rather than what *someone else* has to say—humans tend to be egocentric. Overcoming that tendency takes energy. We have to turn off the commentator in our heads, who is preparing what to say when the next pause occurs, and instead focus on the other person who is talking (or signing). Instead of listening to respond, we must shift to listening to understand. That requires effort, because to understand

★ ★ ★ ★

means we have to think about, process, and weigh what the other person is saying. When we are listening to respond, we already know what we think, which doesn't demand mental energy.

There are specific methods and training programs to help people learn to listen more effectively. Undoubtedly, the most familiar method used in leadership development training is the concept of active listening. While a valuable technique, active listening often gets distilled into a simple process of listening with intermittent episodes of paraphrasing and repeating back what the speaker said. When we approach listening like that, there's no dedication to actually understanding—instead, we go through a process aimed at *seeming* to be listening. This ends up doing a disservice to the people taking part in the conversation. To bolster this approach, the Center for Creative Leadership has identified six components of effective active listening (Table 11.1). An approach to active listening that incorporates these components not only improves the efficacy of communication (knowledge transfer and understanding), but it also increases satisfaction among those on the receiving end of it.

Table 11.1. Components of Active Listening

Pay attention to the person speaking
Withhold judgment about what they have to say
Reflect back what you think they have said; paraphrase concisely
Clarify your understanding by asking questions
Summarize what the person has said
Share your perspective on what the person has said

★ ★ ★ ★

While active listening as presented in The Center for Creative Leadership's model represents an advancement over simply listening to respond, aiding in demonstrating the listener's attentiveness to what someone is saying, it still has major limitations. As a process, it is focused principally on demonstrating to the speaker that you are paying attention. In essence, it shifts our focus from listening to respond, in a real sense, to listening to repeat. Such a process doesn't take into account the emotional components of what is being said, and it doesn't address meaningful understanding. The steps in this model really only focus on listening for content. While this is better than listening to respond, in the end, real, effective listening is often not accomplished.

An alternate listening practice I have found more effective is called reflective listening. This approach was first described by the renowned psychologist, Carl Rogers, and more recently has been described by Charles Duhigg in his book, *Supercommunicators*. Reflective listening involves asking three questions. First, "What is the topic here? What is being said?" Next, ask, "How does the speaker feel about this topic?" Finally, "What meaning does this have for the speaker?" While I feel this method is stronger than active listening, as I've considered it, I have found there is another key question to add—"What is the speaker's understanding of the topic?" I would argue that the speaker's understanding of the topic deeply affects the other factors, and it affects the process of listening. I listen differently to a five-year-old talk about space than I listen to Carl Sagan. In my experience, considering the speaker's understanding of the topic is a helpful component of listening to understand. Figure 11.2 presents these four elements of powerful listening and the questions to keep in mind regarding them. I have found that using this process can greatly improve what I get out of listening.

★ ★ ★ ★

THE FOUR ELEMENTS OF
POWERFUL LISTENING

MEANING
What does the topic mean to the speaker?

CONTENT
What is the speaker's topic?

FEELINGS
What does the speaker feel about the topic?

UNDER-STANDING
What is the speaker's understanding of the topic?

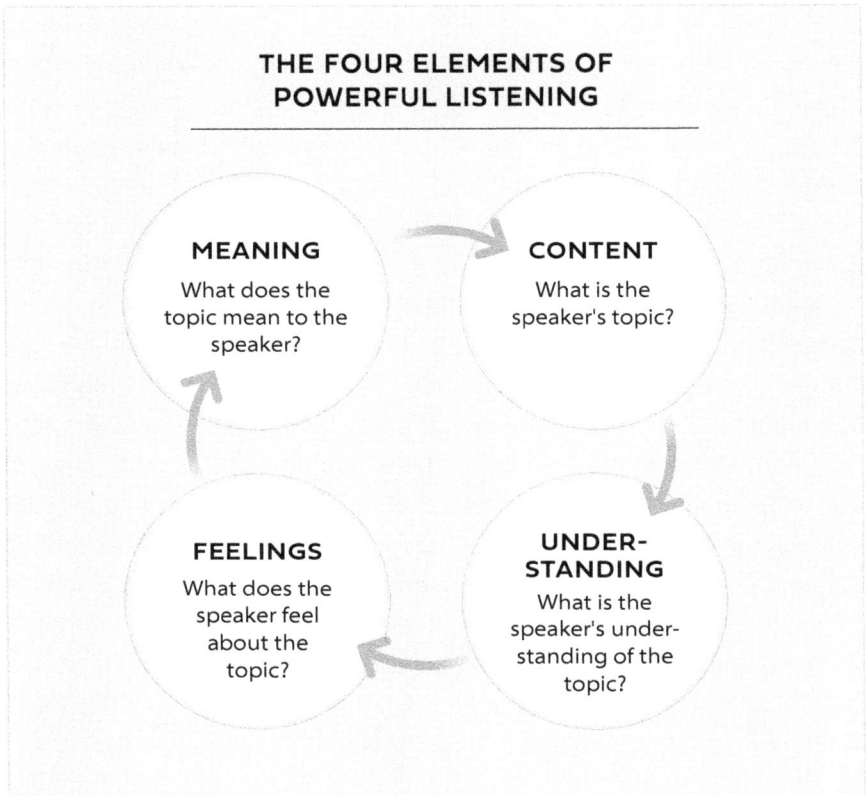

Figure 11.2. The Four Elements of Powerful Listening. To listen effectively, you must identify four things: the content of the speaker's message, their understanding of the topic, their feelings about it, and what it means to them.

Structured listening approaches, whether active listening, reflective listening, identifying the four foci of a conversation, or some other approach, are directed at helping the listener understand what the speaker is really trying to communicate. Four-Star leaders know communication is as much about listening—and understanding—as it is about speaking. General Dave Rodriguez spoke at length about this.

> When I talk about communicating, I talk about listening. We've got to be able to listen very, very effectively. I laugh and tell everybody all the time I've got two ears and one mouth. If I were using that ratio, I'd be a lot better off. You've got to be able to listen, and you've got to listen empathetically to people. You've got to give everybody a chance to get

★ ★ ★ ★

their say, and you've got to spend some time doing it. There's a lot of wise people out there and [...] it'd be good to listen to them as much and as often as you can.

Four-Star leaders listen to hear, to learn, and to understand, not to have the opportunity to respond and demonstrate their knowledge.

SELF REFLECTION

Do you listen twice as much as you speak?
What would be the impact if you spent two-thirds of the time in your conversations listening with a willingness to be persuaded by the other people's ideas?

HOW CAN WE LISTEN WITH A WILLINGNESS TO BE PERSUADED?

Practices like active or reflective listening can greatly improve our ability to listen and communicate, and any leader can benefit from knowing how to use those tools. These practices can be made even more effective by understanding and mastering the powerful listening questions just covered. Yet, these practices and techniques don't get at the fundamental drivers that prevent us from actually listening.

The reason we fail to listen to other people is much deeper than our lack of a process—it derives from our own insecurities and the assumptions we make about others. We have a psychological need to maintain our self-image as smart or competent—as right. When someone else's thoughts or perspectives threaten our self-image and desire to be right, we respond negatively. Most often, we feel a compulsion to defend against any evidence that could prove us wrong, and argue, either verbally or in our heads, that any perspective different from ours is flawed. When we do that, we stop listening and shut off any possibility of being persuaded, leading to an arrest of meaningful

★ ★ ★ ★

communication. If we are to overcome those tendencies and maintain clear lines of communication, we have to take a different approach.

Addressing our natural tendencies to shut down communication requires a complete reorientation of our mindset, and doing so may require a great deal of personal work. It almost certainly will require looking at ourselves in the mirror and admitting we have shortcomings. Luckily, when we do that work, we will become more able to be open and receptive to what others have to say. Consequently, we will be able to fully access the knowledge and expertise of others that we need to lead successfully. To change our mindset to be more open to listening to others requires the development of the virtues of character: humility, honesty, respect for others, and selflessness (see Section I of the book for more details). By developing those virtues and changing your mindset, you will be ready to listen with a willingness to be persuaded.

When General Jim Mattis said that phrase to me—"Listen with a willingness to be persuaded"—it was something I had never heard, and yet, I immediately understood that it held great potential to change how we communicate. Over the ensuing months, as I interviewed one Four-Star after another, I got glimpses of how they listened and what seemed to help them do so. When I coupled all of those experiences with my training to be an executive leadership coach, I began to develop a set of precepts to guide me toward how to listen with a willingness to be persuaded. I have come to see those precepts as seven interdependent rules. When the seven rules are followed, they can help reorient our thinking. The rules are presented in Table 11.2 for ease of reference, and the next few pages will cover each in more detail.

★★★★

Table 11.2. The Seven Rules for How to Listen with a Willingness to be Persuaded

THE SEVEN RULES FOR HOW TO LISTEN WITH A WILLINGNESS TO BE PERSUADED*
1. Relinquish the need to be right
2. Assume the other person is intelligent, thoughtful, and sincere
3. Recognize that the other person has the perspective they do because they have information about it that you don't
4. Suspend judgment on what they have to say
5. Avoid zero-sum thinking
6. Accept that just because they may gain from being right, that doesn't mean they aren't right
7. Recognize that just because they might be wrong, that doesn't mean you're right

* The rules are presented in no particular order.

Relinquish the need to be right

Is there anyone out there who wants to be wrong? Most people think of themselves as smart and thoughtful. So, when we form an opinion or hold a perspective on something, we are fairly certain we are right. If we are proven wrong, it can hurt our ego and self-image. We have a psychological need to be right. Consequently, we often hold tightly to our view, even when it's wrong.

Listening with a willingness to be persuaded requires being open to alternative views and information. It demands we suspend our need to be right—that need to preserve our perspective and self-image—in favor of seeking a better understanding and perspective. Letting go of our need to be right also relieves us of worrying about getting credit for being right. According to General Gus Perna, not worrying about getting credit is a marker of great leaders.

★ ★ ★ ★

> The best leaders are the ones that know how to use all capabilities and capacities around them to get the mission done. [...] They fundamentally believe that a group of people or organizations can get more done if nobody's worried about who gets the credit. They build and lead teams to accomplish their purpose without worrying about who gets credit.

Openness, receptivity, and personal humility are the keys to relinquishing the need to be right. By maintaining that mindset, Four-Star leaders can shift their focus from proving they are right to listening to and understanding others' perspectives, allowing them to broaden their knowledge and perspective in the process. When we broaden our perspective and increase our knowledge, we are in a much better position to lead our people, make better decisions, and have better outcomes.

Assume the other person is intelligent, thoughtful, and sincere

It's not uncommon for people in leadership positions to assume incorrectly that they are the smartest and most dedicated person in the room—any room. Consequently, it can be easy for them to assume their ideas and perspectives are better than those of others. In fact, many leaders have the uncanny ability to believe, despite the mathematical improbability, that they always have the best perspective and solution regardless of the abilities of those they lead. Unwittingly, we can cut ourselves off from valuable insight and information because when we think people aren't as smart as we are, we treat them that way, and, amazingly, they notice. No one wants to be treated like they are stupid, so as a result of our attitude, they avoid us and don't share the information and perspective we need. That is a deadly situation for leaders.

If Albert Einstein were alive and showed up in someone's office, most people would have the sense to listen to what he had to say, no matter what the subject. We all know Albert Einstein as intelligent, thoughtful, and sincere, so we would listen attentively. Unfortunately, we may not be prone to afford others the same respect and courtesy, and that is a disservice to ourselves. By adhering to the other six rules, Four-Star listeners are able to assume the person is intelligent, thoughtful,

★★★★

and sincere. If the listener assumes this to be true, it suggests that the speaker has carefully considered their words, allowing the listener to benefit from their intelligence and perspective. This fosters more opportunities for learning and is distinctly different from what happens when we don't think of people in this way. "One of the keys [to effective leadership]," according to General Ed Eberhart, "is that you have to listen; you have to pay attention to the experts who are on your team." The best leaders see their people as intelligent, thoughtful, and sincere, and they listen to them to make the best possible decisions and accomplish the mission.

Recognize the other person has the perspective they do because they have information you don't

In any complex situation, no individual has the complete picture. General Bob Kehler expressed this idea when he said, "You don't have all the information you need. Somebody else has information out there [that you need]." He's right. As humans, we have limitations and biases that prevent us from seeing things from every possible perspective. By definition, we can only see things from our own perspective, and that is limited. The only way to expand our perspective is to incorporate the perspectives of others. General Bob Magnus likened our perspective to being mentally confined. He said, "The idea is to try to look at things from different perspectives, not just your own [...] because that's almost like a box. But to look at it from other people's perspective to learn."

Four-Star listeners recognize the need to broaden their perspective by learning from others and understanding that when someone else has a different perspective than they do, it is because the other person must have additional information that the listener doesn't. Recognizing this can also help mitigate the need to be right because it helps you see that one's perspective and what they hold to be right is contingent upon the information they have and their understanding or interpretation of it. By approaching conversations with the mindset that people are intelligent and have information we don't, we can greatly improve our ability to listen with a willingness to be persuaded.

★ ★ ★ ★

Suspend judgment

When we listen to someone else speaking, it can be easy to lapse into assessing the validity and veracity of what they are saying. Afterall, we don't want to agree with or even listen to someone broadcasting nonsense. If we enter a conversation without recognizing the other person's intelligence or considering that they may have knowledge we lack, we risk dismissing their perspective while failing to listen fully. We can latch onto one thing that we consider incorrect and totally miss or dismiss the rest of what they have to say. If we do that, there is a good chance we will miss valuable information that could help us. Four-Star leaders suspend their judgment of what's being said and allow the other person to complete their thoughts. That suspension of judgment is at the heart of listening with a willingness to be persuaded. In doing so, we can avoid drawing premature conclusions, making assumptions about what the speaker might be saying, or discounting the speaker completely.

Avoid zero-sum thinking

Many of us, especially in highly competitive fields, see things as a competition where there are winners and losers. This is known as zero-sum thinking, the notion that if one person is to win, the other must lose in order to balance out some sort of hidden success equation. It reflects a win-lose philosophy, or what James Carse referred to as a "finite game" in his book *Finite and Infinite Games*. Contrast the finite game with the infinite game, which is one where everyone can win—there's enough potential success for everybody. Additionally, by definition, infinite games continue as long as players choose to 'play' because opportunities remain for everyone to keep progressing. In fact, in an infinite game, the success of others contributes to your success. Consequently, infinite games are fostered by seeking common ground and cooperation instead of competition. By fostering a win-win approach to problem solving, we promote a collaborative process that allows everyone to be successful and to "keep playing."

★★★★

Accept that just because they may gain from their perspective, that doesn't mean they're wrong

When we have a win-lose mindset, we usually perceive that the other person has something to gain from being right, and we have something to lose if they are. As a result, we may tend to discount what they have to say. This is because it can be easy to see them as biased by their potential gain. (Ironically, we fail to see our own opposing bias.) Four-Star leaders seek to be fair and objective, weighing what is said based on its own merit. They know that just because someone has the potential to benefit from a situation, their perspective isn't automatically invalid.

Recognize that while they may be wrong, you're not necessarily right

We are all susceptible to the cognitive bias known as the fallacy of the false dilemma. This fallacy arises from the incorrect assumption that if one position is wrong, then the alternative must be correct. We see this constantly in the positions of various political figures and pundits. Complex issues get oversimplified into just two options, ignoring the potential for many others. The reality is that both sides could be wrong, and there may well be a completely different position or solution that is correct.

Political figures and pundits aren't the only ones who fall prey to the fallacy of the false dilemma. Most of us do. When confronted with information, a perspective, or an argument that is different from ours, we often do all we can to disprove it. That's the fallacy. In our minds, if we disprove the opposing argument it will mean that we are right, and our ego will be preserved. But that is simplistic and irrational. While the other person may be wrong, you're not necessarily right. In fact, given the number of possible solutions, accounting for all information, perspectives, and arguments, the mathematical probability is that you're more likely to be wrong. Nevertheless, we continue to be victims of the fallacy of the false dilemma, and that threatens our ability to actually get to the right answer.

★★★★

Four-Star leaders don't allow themselves to fall into this fallacy. They avoid it by being open to the possibility that they may be wrong. A major part of that is being able to set aside their need to be right and instead focus on getting to the right answer, and, as General Perna noted, to do so "without worrying about who gets the credit." When we are open to the possibility that we, as well as everyone else, may be wrong, we are able to carefully and meaningfully consider others' ideas and perspectives. That practice of considering varying ideas and multiple perspectives significantly increases our likelihood of getting to the best answer.

★★★★

CHAPTER 12

IT'S NOT WHAT YOU SAY;
IT'S WHAT THEY HEAR

*What I learned was, when you're talking to people […] if they don't speak
the same language you speak, [and you're] trying to tell them what you
want […] you've got to be damn sure they understand. It's not what you
say that's important; it's what was heard that's important.*

– ADMIRAL LEIGHTON "SNUFFY" SMITH, U.S. NAVY –

When he was a colonel, Gus Perna was preparing to transfer to another duty station, and the brigade he was leaving hosted a farewell event for him. As part of the festivities, then-Colonel Perna gave a farewell speech to his soldiers. With over two decades of leadership as an Army officer, as well as extensive training in communication, and numerous experiences giving addresses, Colonel Perna felt he was a polished communicator, and he was pleased with how his speech had gone. That is, until his wife, Susan, shared her perspective.

Setting the stage, General Perna said, "My wife is a social worker by trade, and she's always been my best sounding board [because] she'll tell the king he's not wearing clothes—always—and she has no fear." He added, "She's wicked smart, and she's pretty direct." Because of her astute observation and forthrightness, General Perna had become accustomed to feedback from his wife. So, it was no surprise when she shared some with him as he exited the stage.

★★★★★★★

"Gus, I know you. I watched you," she said. "You thought that was one of the greatest pep talks you ever gave." Instead, she reported, "They have no clue what you were talking about—none. *None.*" She continued, "Either you were ahead of your own thoughts, or you think they're in a different place." As she was saying this, he replied, "I was watching the room, [...] you were looking at the wives," to which Susan responded, "Oh, no. Oh, no. I was watching your XO (executive officer). I was watching your commanders." Then, she said "one of the strongest things [General Perna has] ever learned, and [something he has] always kept in mind: 'Gus, it doesn't matter what you say; it's how they receive it.'"

Reflecting on that communication maxim, General Perna said, "Words matter. Body language matters, and, as Susan said, perception matters." He surmised, if you have "ten people in a room, ten people are gonna [...] hear or view it differently. You really have to become a master of [communication]."

What does it mean to communicate? As leaders, it can be easy to think of communication as our transmittance of information to the led. Afterall, we see Presidents giving speeches and CEOs making announcements. Isn't communicating the mission, vision, and values what a leader does? A leader stands somewhere and transmits the information, the plan, the mission to the led, and when we as leaders do this, we think that communication has occurred. General Perna's experience exemplifies a reality I've covered earlier: communication is about much more than transmitting information to someone else.

Over three-quarters of the Four-Stars agreed that communication isn't nearly as much about what you say as it is about what the receiver hears and understands. That is a different way of thinking about communication than most people are used to. I have come to believe it is the only way to think about communication, and it has led me to ask myself questions. How can I shape my communication so that it accomplishes the intended goal—to ensure that my audience hears and understands the actual message I hope to convey? How can I master what some of the Four-Stars suggested is the *sine qua non* of leadership?

★★★★

As I have with so many other concepts in this book, when I began thinking about how to be the most effective communicator, I started thinking about a systematic process that would capture not only the components of communication, but also the barriers that prevent it. Consequently, I devised a framework of the communication process that, if mastered, can allow us to excel at what may be the most complex part of leadership—clear and effective communication.

Before getting to the communication process, you may be wondering why, if communicating effectively is so important to leadership, it hasn't been covered until now. There are a few reasons for that. First, to be the most effective communicators, the kind who convey important and meaningful messages with clarity and caring, we have to have mastered Character, Competence, and Caring. If we haven't, those we are hoping to communicate with won't trust, listen to, or believe us. So, to truly master communication we must have mastered all the concepts in the book to this point. The second reason I have saved this discussion to the end of the book is because the communication process is complex and difficult to get right, let alone master. Whereas most of the concepts covered previously may be difficult, they aren't multifaceted processes with numerous potential failure points; you either understand them, or you don't. The communication process isn't that way. The final reason to save the communication process to the end of the book is to take advantage of the recency effect—the cognitive bias that makes us more likely to remember the information presented more recently than that which was presented earlier. That is, this is an attempt to help you remember this more complex and highly important leadership concept.

WHAT DOES IT TAKE TO COMMUNICATE?

Communication is the process of sharing ideas and information among people. While it can be tempting to think of it as a two-step event composed of one person transmitting and another receiving a given message, a leader does that at their own peril. As Figure 12.1 shows, the process of communicating is much more complex, comprising at least

★ ★ ★ ★

four principal steps and three influencing determinants (Communication experts may identify more). To break down the process, it may be helpful to consider first the steps and then the determinants.

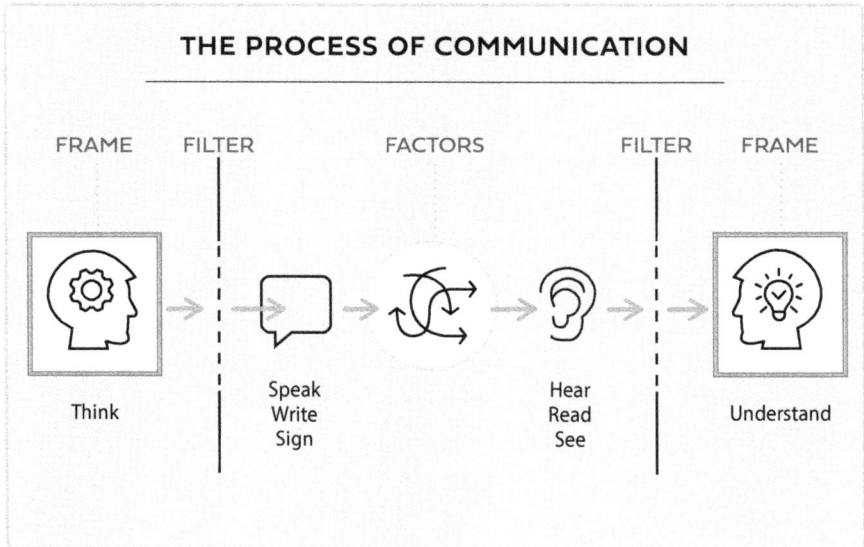

Figure 12.1. The Process of Communication.
Before communicating, we think about the message, and we do so within the Frame of our knowledge and prior experiences. We Filter that message through the relationship we have with those to whom we then Speak, Write, or Sign. Other Factors, such as the topic, context, and nonverbal communication influence the message the recipient Hears, Reads, or Sees. The recipient Filters what they hear through their prior relationship experience with us to receive a message they then consider and understand within the Frame of their own knowledge and prior experiences.

The Steps in Communication

There are four steps in communication: Think, Speak, Hear, and Understand. The first two steps in communicating a message are on the part of the person transmitting the information. The speaker can control those two things (i.e., think and speak). The latter two (i.e., hear and understand) are in the domain of the person receiving the information. While the speaker cannot control these two, they should inform how the speaker conveys the message.

★★★★

It seems obvious that before you communicate something, you have to think about what you want to say. Yet, there are innumerable examples of people who do not seem to think before they speak. Four-Star leaders always do. In discussing what he described as the "hugely important" role of effective communication in leadership, General David Rodriguez said, "For the most part, [leaders are] undisciplined at how we communicate. We don't think about the impact that communication is having on people. [...] You've really got to think about what you're going to say before you say it."

After thinking about what to say, you then have to say (or write or sign) it to your intended recipient(s). But there's more to it than just that transmission. To be effective, you have to consider your recipient. General Rodriguez said, "You've got to be able to speak so that they can understand, otherwise you're not communicating." Failing to consider our audience is perhaps the most common pitfall leading to miscommunication. When we don't consider our audience, we are highly likely to deliver the message in a way that the audience cannot understand.

Let's conduct an illustrative thought experiment by considering John F. Kennedy's inaugural address. There's no doubt he must've thought long about what he would say. He spoke eloquently and in ways that the American people could understand. The most famous portion of the speech—"Ask not what your country can do for you; ask what you can do for your country"—is a powerful admonition almost everyone listening on the Mall in Washington, DC could understand. But, what if instead of speaking English he delivered his speech in Icelandic? Only a tiny fraction of those who heard his address would have any idea what he was saying; he would have not communicated at all, even though he may have been transmitting wonderful ideas in the most beautiful Icelandic. You have to speak in a language—in a way, in a form, in a manner—that your intended audience can understand. If you don't, failure in communicating is nearly guaranteed.

Almost everyone, except perhaps those who have difficulties with sensing and responding appropriately to social cues, recognizes the need to adjust the way we communicate to fit our audience. Specifically, when communicating with children, most adults automatically adjust

★ ★ ★ ★

what they say to be appropriate to the children's developmental level. In fact, you can even see older children adjust their communication to be understandable to younger children. This ability to recognize and adjust our communication to fit our audience seems innate. Yet, as adults, it can be easy to forget the value of doing so. If we want to have the best chance of communicating effectively, we have to adjust our communication according to the characteristics of our audience. When we don't, we introduce opportunities for miscommunication.

Following the transmittance of the information, the recipient hears (or reads or views) the message and processes it to understand it. Though you, as the speaker, do not directly control the last two steps in the communication process, you can improve the likelihood that the intended message will be understood. Consider how your recipient may hear and understand the particular message. Based on what you know of the recipient, how might they hear and understand it in a different manner than you intend? What are the ways it could be misinterpreted? After asking these questions, use what you've identified to inform how you think about and transmit the message.

These four steps just covered can be deceptive, appearing as if they are aspects of a seemingly straightforward transmit-and-receive process. Don't be caught off guard. These steps are heavily influenced by three major determinants—Frames, Filters, and Factors. When not considered and used to shape your communication, these three determinants have the potential to completely transform your intended message into something totally unrecognizable. It could be argued their impact on communication may be greater than the actual message itself.

WHAT ARE THE THINGS THAT SHAPE OUR COMMUNICATION?

Mental Frames

If we conceive of communication as a straightforward conversation between two people, each person comes to that conversation with their own set of prior experiences, knowledge, and cognitive biases

★★★★

that frame the way they see and interpret the world. This could be referred to as their perspective or frame of reference. But, for the sake of the discussion here, I will refer to it as their "frame."

Our frames inform how we compose and send, as well as receive and interpret, communication. They shape how we think about the topic of our message, including why we feel the topic is important, and what we want to share. In reality, without the prior knowledge and experiences which create our mental frames, we probably wouldn't have a message to communicate. On the other side of the conversation, our recipient also has a mental frame shaped by their knowledge and life experiences. Their mental frame shapes how they respond to, think about, and value what we are trying to communicate. Their frame creates a set of expectations, which focuses them on certain parts of a message, and blinds them to other parts, so that they might come away with an incomplete or distorted understanding of our intended message. In this way, people can leave a conversation having heard what they wanted to hear.

General David Rodriguez recalled various times when he was "in a big room with a lot of senior people" who were discussing high-level topics. After someone had communicated a mission or plan, he witnessed multiple times when, "Somebody would say, 'Okay, this is the way it's going to be,'" which clearly revealed that person had not heard the message that was communicated. Likening it to a game of "telephone," General Rodriguez said, "The people heard what they wanted to hear, not what was said." This can obviously have significant implications for the mission and success of a given group or organization, and we must mitigate that. To do so requires some understanding and recognition of people's cognitive biases.

Cognitive biases that degrade communication

Hearing what we want to hear is a common human phenomenon that can occur in essentially any setting where people are trying to communicate. There are multiple cognitive biases that drive this tendency, and almost everyone has them. When we gain knowledge of those biases, we can think through the way we deliver communication, thereby

★ ★ ★ ★

increasing the potential for our recipient(s) to hear the message we are trying to send. While there are far too many cognitive biases to cover sufficiently here, let's touch on two that often lead to us hearing what we want to hear.

It is important to remember that most people are wired to avoid pain, both physical and mental. From a mental standpoint, we generally avoid things that induce anxiety or psychological discomfort. This includes when our self-image is threatened. Imagine that you are someone who thinks of yourself as intelligent and highly capable. You have years of successful life experience that confirm that self-image. Now, imagine that you do something that begins to fail. As the failure unfolds, it stands in stark contrast to your self-image, and that contradiction creates psychological discomfort. This is cognitive dissonance—when reality stands in opposition to how we view ourselves and the world around us. To minimize the psychological discomfort caused by cognitive dissonance, our tendency is to ignore the facts and make up alternative explanations for why things have happened, all to protect our self-image. Many of our cognitive biases play a significant part in alleviating cognitive dissonance. Two highly common cognitive biases that distort our ability to hear the message being communicated are *attention bias* and *confirmation bias*. These two biases interact and can have powerful effects on our ability to hear a message as it was intended.

Attention bias is our fundamental tendency to pay attention to information about things we are already thinking about or that interest us. If you like soccer and hate scrapbooking, when someone talks about a soccer match, you will pay attention, whereas you will not pay attention to a discussion on scrapbooking. We pay attention to the information we find interesting and ignore the rest, often becoming completely unaware that the information we ignored was ever discussed. If a leader is speaking about issues A, B, and C, and we are only interested in issue B, we will only "hear" B and will miss A and C altogether. For those of us in leadership positions who frequently present information, this means that we must always consider what is of interest to our audience and how we can shape our message to keep them

★★★★

interested in what we're saying. If a subject is of no interest to your audience, you are unlikely to gain their attention and effectively communicate with them.

Confirmation bias is our tendency to acknowledge and/or search only for the information that aligns with our existing perspective or mental frame. It keeps our egos from being threatened by being wrong about something. This means we pick out the information we want to hear and discount the rest. This is a long-recognized phenomenon. Sir Francis Bacon wrote in 1620 that once we have formed an opinion, we find "all things else to support and agree with it." Voltaire said something similar: "The human brain is a complex organ with the wonderful power of enabling man to find reasons for continuing to believe whatever it is that he wants to believe." Confirmation bias is a major contributor to groupthink and societal polarization—people surround themselves with others who think as they do, and they confirm and amplify each other's beliefs. Meanwhile, they vehemently discount any data that contradict them. You may be familiar with this phenomenon if you have been paying attention to our society, particularly how groups have formed around different political ideologies—one group is certain their perspective and supportive data are correct, while another group has just as much certainty about their own. To break through that rigid defense requires having open and honest dialogue. Doing so can help us understand the other person's perspective, which can then inform our own and how we communicate with them. We tend to be more open to different ideas when we know that we have been listened to and appreciated. While that takes time and effort, it can help us overcome confirmation bias in our recipient(s), which can lead to more effective communication, saving time and effort in the long run.

Our mental frames—how we see ourselves and the world around us—are heavily influenced by our cognitive biases and our tendency to minimize cognitive dissonance. These phenomena, perhaps far more than the actual words someone speaks, have enormous influence on our ability to hear and understand what someone is trying to say. As leaders, if we hope to be effective at communicating, we must

★ ★ ★ ★

recognize these cognitive phenomena and people's mental frames, then shape our communication accordingly.

Before moving onto Filters, take a moment to consider the mental frames that you have that shape the way you think about the world and how you communicate—in good ways, as well as in limiting ways. For instance, on one hand, as someone who grew up with modest resources in a family with limited education, my worldview is framed by those experiences. On the other, as a highly trained physician who has lived on the East and West Coast, as well as in the middle of the country, the way I see things is also framed by those experiences. There are a host of other things that impact how I view the world and how I communicate. The same goes for anyone with whom I hope to communicate, which means I need to know enough about them to take those things into consideration.

Modifying Filters

Beyond our frames, each person in a conversation filters the information transmitted and received through their relationship with the other person(s). The relationships may be professional, transactional, platonic, romantic, familial, or otherwise. After thinking about the message, a speaker uses the relationship they have with the recipient(s) to filter how they convey the message. Whereas someone might speak frankly or colorfully, with more emotion and, perhaps, strong language, about a topic to a close friend, the same person will likely filter the message when speaking to a professional acquaintance or someone in an official position. We typically would describe this as the need to "know your audience," which General Joe Ralston said leaders must do. Within the context of the relationship with an audience, he said, "You may have to modify your communication approach."

Just as the speaker filters what they say through the relationship with the recipient, the recipient filters what they hear through their relationship with the speaker. The strength of the relationship with the speaker has a direct effect on how the recipient filters the message. For instance, think about someone close to you—a friend, spouse, relative, etc. Most of us have had times when we didn't transmit our

★ ★ ★ ★

message perfectly to someone close to us, and yet they heard what we were trying to say—they knew what we meant. Conversely, if you have ever had a difficult relationship with someone, sometimes it can seem like they tried to misunderstand you no matter how meticulous you were in trying to deliver the message. For better or worse, all of our communication is filtered through the relationships we have with our audience. To be effective communicators, we must always keep those filters in mind.

SELF REFLECTION

How consistently do you consider how your message needs to be modified so that your target audience will hear what you intend to communicate?

Think of a recent example where someone misunderstood what you were trying to communicate. How could you have communicated your message more effectively?

General Darren McDew, *on Filtering Your Communication to Your Audience*

As a highly performing Air Force pilot, then-Captain Darren McDew was selected to be an instructor in the Formal Training Unit (a pilot-training program) where the instructors were "top of the top" pilots in their particular aircraft. All of the program instructors were at the rank of captain, with the promotion boards for major coming soon.

One weekend, the instructors got together at a park on base to have a cookout. As they sat around the picnic tables talking, the topic of the promotion boards came up. General McDew recalled, "We're all coming up for major about this time, and people [were] fretting because the promotion rates for major [were not high, and] a lot of people didn't make major." As they sat at the tables having burgers and beers, Captain McDew listened to the other instructors worrying over

★ ★ ★ ★

the issue. In that more socially oriented situation with his colleagues, he shared his perspective on the promotion boards.

> I said, "There are certain things you're responsible for getting done. You can control your performance. You can get your education done; you can get your PME (professional military education) done. Then after that, I'm not worried about [it]. I'm just going to put [my] all into those things. That's all I can control. So, I'm not worried about anything else. I'm just worried about what I can control."

Unfortunately for Captain McDew, some of the other instructors interpreted his pragmatic and frank statements differently than he had intended. He said, "What they heard was 'Darren McDew is not worried about making major.'" As a result, some of his peers and others labeled him as being overly confident and sure of himself. General McDew reflected on the experience and the valuable lesson he learned. "My hard lesson there was, 'Be careful how frank you are in what crowd,' [...] Now, I didn't change [my willingness to be frank]. I just now understand how. I'm still frank in every crowd, but I'm now careful. I didn't understand what could come out of this when I'm that frank." General McDew explained how he would've done it differently now.

> If I'd have been that aware at that time of my life—and I wasn't... I could have easily said [...], "Hey, we're all sweating this thing." [That would] get total empathy out there. "I'm concerned, [too]. I don't know if a year or two from now I'm going to be a major or still a captain, but what I'm focused on right now is the things I can control." Now, that's a whole different conversation—everybody hears it.

Muddling Factors

The final determinant of communication are the specific factors that shape how our recipient hears the message we transmit. These factors can include numerous things, but the most common are the topic, context or environment, and our nonverbal signals. As leaders, we have to recognize and mitigate these factors if we want to be successful at allowing our recipient to receive and understand our message.

The *topic* can have a significant influence on how we communicate, whether sending or receiving a message. Different topics can

★ ★ ★ ★

produce varying emotional responses. Depending whether the topic is a performance review, a complaint, a discussion of policy changes, or catching up on vacation plans, the parties communicating may be anxious, defensive, bored, or excited, respectively (or any number of other responses). Our emotions have direct and profound effects on how we both think about and send messages, as well as how we receive and interpret those from others. They impact our ability to pay attention, reason, and learn—all of which are necessary for effective communication. When we don't recognize the impacts the topic can have on our ability to think about, shape, and transmit the message, as well as our recipient's ability to hear, interpret, and understand the intended message, we are almost guaranteeing miscommunication will occur.

The *context* or *setting* in which we are communicating can also significantly impact communication. In a loud environment, someone who is yelling to be heard may be perceived as being angry, which can shape how the recipient receives and interprets the message. If parts of a message are difficult to hear, whether from ambient noise or poor connectivity via telephone or internet networks, the recipient may try to infer what the complete message is, which can lead to dramatic differences between what is intended to be communicated and what is understood by the recipient. If the communication happens in a social setting, such as with General McDew's earlier story, one party may view the message from a social context (i.e., just two friends at a cookout), while the other may view it from a professional perspective (i.e., two pilots competing for promotion), resulting in miscommunication on the part of each. As leaders, we must consider how the context can affect our communication and adjust accordingly.

Among the factors that serve as determinants to the effectiveness of communication, perhaps none is more commonly emphasized than nonverbal signals. As covered briefly in Chapter 10, it is well known that in many cases our nonverbal signals can impact communication as much as the words being transmitted. Expanding on what was covered previously, nonverbal signals include a number of different things that have a powerful influence on the message being received:

★ ★ ★ ★

gestures, how we stand, our tone of voice, our volume and cadence of speech, whether and how much eye contact we make, where we stand or sit relative to the person to whom we are speaking, if we smile or frown, how we position our hands and arms, and whether we use physical touch. Nonverbal signals help our recipient(s) know how important the message is to us and how engaged they should be with it. When what we are saying and our nonverbal cues are aligned, our recipient(s) can process those messages together to arrive at a more accurate understanding of what we are trying to communicate. Our recipients will also tend to view us as more honest and trustworthy. However, when there is dissonance between our verbal and nonverbal communication, we cloud our recipient's understanding of our message. Instead of being able to focus on a clear message, they spend a lot of effort trying to determine how much they should pay attention to either the verbal or nonverbal signals; it's confusing. Depending on the context of the communication, studies suggest that nonverbal signals, excluding vocal features like tone, account for over half of our communication. Given this, when our nonverbal signals do not align with what we are saying, people tend to put more value on the nonverbals, which can derail communication.

HOW CAN WE ASSURE OUR AUDIENCE RECEIVES OUR INTENDED MESSAGE?

Considering the complexity of the steps and determinants in the communication process, how can we ever expect to be effective at communicating? While the previous section covered large concepts to consider in shaping how you communicate a message, among the Four-Stars, two fundamental principles of communication stood above all others: simplicity and clarity. In fact, most of them emphasized the absolute need to keep communication simple and clear. For instance, General Gene Renuart said it is "important whether you're talking to a small team or you're sending out an operations order [...] to 150,000 people." He said when a leader communicates a message, it needs "to be as clear and as concise, yet specific, as you can [make it] so that the

★★★★

message gets across without a lot of confused looks on people's faces." A part of keeping the message clear and concise is making certain there is nothing in it that requires interpretation. When our messages require interpretation because they are complex and unclear, people will instead rely on their frames, filters, and factors to determine the meaning. This inevitably leads to miscommunication. When that happens, those we lead often end up doing things we don't want. General Skip Sharp described this, saying, "I firmly believe, throughout my career, that if there's a failure of a subordinate—he does something wrong or just goes completely off on a tangent that's not in concert with the mission of the organization—the great majority of the time it's because he hasn't had enough guidance or explanation..." A Four-Star communicator will simplify things and say, "Okay, here's what the mission is, and here's what we're trying to do."

Admiral Leighton "Snuffy" Smith, *on Making Instructions Simple*

As the commander of the aircraft carrier *USS America*, then-Commander Snuffy Smith made it his practice, after flight operations ended at 0100, to "get up and walk around the ship at night." He would "go around the engineering spaces" so that he could "talk to the guys in engineering [and] learn a little bit about them." He found "There's a million things you can learn about a ship when you walk into somebody's space at midnight, and you start talking to them."

One particular night, Commander Smith was walking about the ship and "went down to the absolute bottom of the ship in the engineering spaces." There, in the sweltering bowels of the aircraft carrier were six enormous boilers that weighed hundreds of thousands of pounds and stood over 20-feet-tall. Powering the ship, each burned at approximately 2,500° F and was at risk of overheating and causing a catastrophic explosion. On the front of each boiler was a glass sight-gauge where the boiler flame could be seen, and there was a glass tube that showed the amount of water in the boiler. A young sailor was stationed at the front of each boiler to monitor the flame in the sight-gauge and the water in the glass tube. If the flame went out or the amount of water

★ ★ ★ ★

in the tube began to decrease, the young sailor had to pull a chain that would shut down the boiler, preventing the boiler from exploding.

Commander Smith began talking with a 19-year-old sailor who was on duty monitoring his particular boiler.

> I said, "Tell me, kid. Why are you here?" He said, "I'm watching the flame. I'm watching this water gauge." I said, "What do you do?" [...] "Well, I pull this chain [and] shut it off [if there's a problem]." I said, "Why do you do that?" [The sailor replied], "The chief told me." I said, "No, no. Why do you do it?"

Recognizing that the dutiful sailor did not understand what he was really asking, Commander Smith reiterated the question. He was hoping the sailor would tell him the underlying reason for what he was doing—"If we run out of water, it's going to blow up." Since the sailor had not understood the question, Commander Smith rephrased it.

> "Why do you do this?" [Again the sailor replied], "Because the chief told me to do it." I said, "Well, I know the chief told you to do it, but why do you pull that chain? What happens if you don't pull that chain?" [Thinking quickly, the young sailor responded], "Chief's going to kick my ass because he's a mean son-of-a-bitch!"

Laughing, Admiral Smith recalled thinking, "That's good leadership!" Recognizing some limitations in the young sailor, the chief had communicated to the sailor in very simple and clear terms what his duty was. Given the gravity of the consequences if the sailor did not carry out his job, the chief had left no room for misinterpretation and, as a result, the young sailor knew what he had to do and was intently dedicated to doing it. Admiral Smith said, "I loved it!"

The need for simple and clear communication isn't limited to those who are inexperienced or naïve; everyone benefits from simple and clear communication. Unfortunately, because our mental frames derive from our knowledge and experience, we often attribute the same level of knowledge and experience to other people. Consequently, we may share a lot of higher order information or leave out required explanations because we think we are keeping things simple and clear. In either situation, we are inadvertently making things complex for the other person. In reality, no matter how experienced or intelligent

★★★★

someone may be, they don't share the same background experiences and knowledge we do—their frames are different from ours. When we assume people have all the information we have, we are at high risk for leaving out vital information from our communication, which almost always leads to misunderstanding.

When we are in leadership positions, it may be tempting to complicate our communication in hopes that it gives us *gravitas* or makes us seem deserving of our position. We might be inclined to think if we communicate in complex ways, people will see us as smart, and that will lead them to view us as better leaders. However, the truth is the exact opposite. When we communicate things simply and clearly, people then know what is going on and what is expected of them. Consequently, they perform better, and the mission is more likely to be accomplished. That is excellent leadership. Woody Guthrie is quoted as having said, "Any fool can make something complicated; it takes a genius to make it simple." Leadership genius is built on communicating things simply, clearly, and in a manner so that what you say is what people hear. However, as the word genius implies, leaders with the skill to communicate simply and clearly are rare. Consequently, if you develop that skill, not only will you demonstrate excellent leadership, but you will also be given greater leadership opportunities.

General Keith Alexander, *on The Impact of Communicating Simply*

In 2004, then-Lieutenant General Keith Alexander was serving as the Deputy Chief of Staff for Intelligence of the Army. At the time, it had come to light that there had been serious events of prisoner abuse committed by Army soldiers and Central Intelligence Agency officers at Abu Ghraib Prison. Because of his position as an Intelligence Officer, Lieutenant General Alexander was tasked by Secretary of Defense Donald Rumsfeld to testify before the Senate Armed Services Committee regarding the situation at Abu Ghraib. Given the seriousness of the offenses and their international impact, the senators were keenly interested in knowing which senior leader should be held accountable for the maltreatment of the prisoners.

★★★★

> I was asked a question by [Senator Hillary Clinton] who was up in the right-hand side: "Shouldn't it be the Army intelligence guy who goes to jail over this, because he's the garrison commander for Abu Ghraib?" I looked at her and said, "Well, you're familiar with Fort Drum (an Army base in her congressional district). At Fort Drum, the garrison commander's an O-6 (colonel). He runs the garrison (the facility where the soldiers are stationed), and the division commander, who runs 10th Mountain Division, runs the division (the group of soldiers). The garrison commander doesn't tell the division commander what to do. He doesn't have operational control of [the division]. He runs the garrison, and the division commander runs the actual division itself. At Abu Ghraib, it's the same thing. The intel officer was responsible for the garrison of Abu Ghraib, but the MP (military police) ran the prison, and that's how that worked." She said, "Nobody explained it to me like that. Thank you."

Lieutenant General Alexander was able to explain in a clear and simple manner the complex organizational structure at Abu Ghraib by using an analogous structure Senator Clinton was familiar with in her state of New York. Following his testimony in the Senate hearing, Lieutenant General Alexander returned to his office at the Pentagon. A short time later, he received a call from Secretary of Defense Donald Rumsfeld who was brief: "Come up to my office." As he put down the phone and rushed off to Secretary Rumsfeld's office, he was wondering what was about to happen. "I thought, 'What did I say? There's so many things that could have been taken wrong.'" When he arrived in the office, Rumsfeld said, "'I just got a call from Senator Clinton. She said you're the only one they understood."

As a result of Clinton's positive feedback, Secretary Rumsfeld told Lieutenant General Alexander, "With former Congressman, Pete Geren, you're going to represent the Defense Department in all the Abu Ghraib briefings for the next four months."

His ability to explain things simply and clearly to Senator Clinton and the Senate Armed Services Committee opened further opportunities for Lieutenant General Alexander to demonstrate his capabilities while interacting with high-ranking military and governmental officials. General Alexander's performance in this delicate and complex situation undoubtedly contributed to Secretary Rumsfeld's decision

★★★★

to appoint him as Director of the National Security Agency the following year.

The world is saturated with muddled messages bombarding us from all directions. It's enough to leave anyone disoriented. Four-Star leaders understand, study, and master communicating clearly and concisely. Consequently, what they say and what others hear is aligned, no matter their audience's rank or station in life, and that is of utmost value in ensuring their ability to influence others to achieve a common goal—the fundamental definition of leadership.

★ ★ ★ ★

CONCLUSION

Every day, at the end of the day, you're a better person, or you're worse, and it's all up to you.

– ADMIRAL JIM HOGG, U.S. NAVY –

On 20 June 2022, I was anxiously awaiting 3 PM. That was when I would begin the first of the 51 interviews that would culminate in *The Four Stars of Leadership*. Not only was it the first interview, but it was also with someone I held in the highest of regard—General Jim Mattis. I knew his reputation. As an officer, he was roundly loved by those he led. As the U.S. Secretary of Defense, he performed admirably, overseeing more than 3,500,000 personnel and an annual budget "well north of $700 billion per year." I had watched how he handled himself with character before Congress and the world. Under great pressure from above, his integrity didn't waiver, and he ultimately resigned his position out of what I believe was a sense of duty. General Mattis' actions as the Secretary of Defense were inspirational to me, and I was about to meet him, at least virtually.

When General Mattis appeared on my screen, I did all I could to maintain composure, while he was immediately disarming and treated me as though we were old friends just having a conversation. When we moved into the structured interview questions, he answered them so thoughtfully and clearly that I felt like I was getting a personal master-class in leadership. The entire time I marveled over the fact that I was getting the opportunity to interview him.

★★★★

Everything was going well, and I was fully engaged and learning a tremendous amount. He had already given me the most memorable quote of the entire project: "There is nothing closer to a god on Earth than a general on a battlefield." I was having an extraordinary time. Then, about halfway through the interview, we reached the second question: What is a leadership lesson you learned the hard way? His answer wrecked me.

In Chapter 3, I recounted General Mattis' answer, when he retold the story of when, as the Commander of the 1st Marine Expeditionary Brigade in Afghanistan, he was ordered not to go after Osama bin Laden. As he told that story, I felt like I was watching him pull back the curtain of world history and letting me see behind it—the real story. I was shocked. I thought of all the Soldiers, Sailors, Airmen, and Marines who subsequently died or were permanently wounded because bin Laden got away. I thought of the civilian casualties and all the political turmoil that followed during the ensuing decade. I was nauseated. I was indignant; not at him, but at the General who had ordered him and his Marines to stand down. Simultaneously, I felt the historic weight of the story.

Reeling from what I had just learned, I was buffeted again when General Mattis said resolutely, "I failed is the bottom line." My internal monologue went into hyperdrive. "What?!?! Did he not hear what he just said? How could he take the blame for that? Isn't it clear who's at fault here? It wasn't him!" Yet, without equivocation, he took the blame, and I was left wondering what to do with it all.

Over the course of several subsequent interviews, I had the opportunity—I felt compelled, really—to share General Mattis' story with other Four-Stars. Many of them knew the story; all appreciated it; and some explained why they agreed with his assessment. For General Mattis and those other Four-Stars who agreed with him, the ability to take ownership of that history-altering event in Afghanistan is one born out of having the character to accept responsibility and accountability as leaders. "In those times when something doesn't work out or something goes wrong," General Jim Jones said, "leaders should [have] enough [character] to stand up and take the blame rather than point

★ ★ ★ ★

blame out to others." General Mattis had done that, and as he pulled back the curtain of history and took the blame, I grew as a leader and a person.

As people have learned about this project, countless have been interested in what it was like to get to interview these exceptional people. They want to know what I have learned. These aren't questions I can adequately answer. I try to convey my deep sense of honor and privilege in having people I hold in the highest regard share their insights with me on a topic that I'm deeply passionate about. I try to share what it is like to spend an hour talking with your heroes. Ultimately, I usually tell them it has been the single most meaningful, formative professional endeavor of my life. But the delightful reality is, my experience with the wisdom of the Four-Stars hasn't ended. I am continuing to learn and grow. I have followed Admiral Jim Loy's admonition that leaders should regularly reflect on their experiences and grow from them.

> There is great value in pausing and looking back over, not maybe just what you did today or what you did last week, but rather where you started and where you came from and how you grew across that span to be a better leader today than you were way back then. [It is] this notion of reviewing and growing.

I have done that, and it is clear to me I am a better person and leader because of this project. It is my intention to continue growing because of it.

Through *The Four Stars of Leadership*, it is my sincere hope that I have been able to convey to you many of the leadership lessons I have learned. In truth, the first draft of this book had twice as many chapters and lessons, but few people would pick up a 500-page leadership book! So, I have chosen only the most fundamental lessons for inclusion, leaving the more advanced lessons for another day. I believe everyone can use the material here to grow in our leadership, with the goal of becoming Four-Star leaders. With that hope in mind, I am reminded of something General Gene Renuart said:

> You shouldn't be someone just willing to accept average. Whether that's talking to your children or it's talking to your team, or it's talking to your

★★★★

boss, let's don't go in[to] this [endeavor] to just tie. Let's go in this to be successful. We can define success in many, many ways. Those can be military; they can be diplomatic; they can be economic; they can be social; they can be cultural. I mean, but don't be the person that just accepts "okay."

As leaders, none of the people we lead deserves for us to be average. They deserve Four-Star leadership—and now you have a toolset to help you provide that for them.

★ ★ ★ ★

BIOGRAPHICAL SKETCHES
OF THE FOUR-STARS

General, U.S. Army (ret.) **Keith Alexander** is a 1974 graduate of the U.S. Military Academy at West Point, later earning master's degrees in business, physics, and systems engineering. Throughout his 40-year military career, Alexander held key intelligence and communications roles. His senior leadership roles included Commander of the Army Intelligence and Security Command (2001-2003) and Deputy Chief of Staff for Intelligence for the U.S. Army (2003-2005). He was then promoted to serve as Director of the National Security Agency (2005–2014) and the first Commander of U.S. Cyber Command (2010–2014), leading 75,000 personnel with an estimated budget of $12B. After retiring Alexander founded IronNet Cybersecurity, advising on cyber defense for governments and businesses.

Admiral, U.S. Coast Guard (ret.) **Thad Allen** is a 1969 graduate of the U.S. Coast Guard Academy, later earning master's degrees in public administration and strategic studies. Allen served in various operational and leadership roles throughout his 39-year career, specializing in maritime security, disaster response, and emergency management, including taking over Hurricane Katrina relief operations, bringing order to the chaotic federal response. He was later appointed the 23rd Commandant of the Coast Guard (2006–2010), commanding 60,000 personnel with an estimated budget of $14B. In 2010, he was appointed National Incident Commander for the BP Deepwater Horizon oil spill, coordinating federal response efforts to contain the environmental disaster. After retiring, he has served on multiple boards of directors and as a senior advisor on emergency preparedness and resilience.

General, U.S. Army (ret.) **Vince Brooks** is a 1980 graduate of the U.S. Military Academy at West Point. He was the first African American to serve as West Point's Cadet First Captain, its highest-ranking student leadership position. During his 38-year military career, Brooks held key command positions across the globe, including in combat in Iraq. His senior leadership roles included serving as Deputy Commanding General, 1st Cavalry Division (2006-2008), Deputy Commanding General, III Corps (2008-2009); Commanding General, 1st Infantry Division (2009-2011), and Commander of U.S. Army Central and Third Army in the Middle East and Central Asia (2011-2013). At the Four-Star level, he served as Commander of U.S. Army Pacific (2013-2016) and subsequently as the triple-hatted Commander of U.S. Forces Korea, United Nations Command, and Korea--U.S. Combined Forces Command (2016-2018) where he commanded 650,000 personnel with an estimated budget of $5B. After retiring from military service, Brooks transitioned to strategic advisory roles, focusing on leadership, international security, business directorship, and defense policy.

General U.S. Army (ret.) **George Casey** is a 1968 graduate of Georgetown University and later earned a master's in international relations. During his 41-year military career, Casey led soldiers at every level, including in combat in Bosnia and Iraq. His senior leadership roles included Commander, 1st Armored Division (1999-2001), 30th Vice Chief of Staff of the Army (2003-2004) and Commanding General of Multi-National Force – Iraq (2004-2007). He was then appointed Chief of Staff of the U.S. Army (2007–2011), leading 1.5 million personnel with an estimated budget of $250B to modernize and rebuild the Army after years of sustained combat operations. After retiring in 2011, Casey has been active in leadership consulting, national security policy discussions, and corporate advisory roles.

General, U.S. Army (ret.) **Pete Chiarelli** is a 1972 graduate of Seattle University and later earned master's degrees in public administration, strategic studies, and national security strategy. Throughout his 39-year career, Chiarelli led soldiers at all levels, including in combat in Iraq. His senior leadership positions included Commander of the 1st Cavalry Division during the Iraq War (2004–2006) and Commander of Multi-National Corps – Iraq (2006). He was then promoted to serve as the 32nd Vice Chief of Staff of the Army (2008-2012), where he helped lead 345,000 personnel with an estimated budget of $240B and became a leading advocate for improving treatment of traumatic brain injuries and post-traumatic stress disorder among soldiers. After retiring in 2012, Chiarelli continued his advocacy as CEO of One Mind, a nonprofit focused on brain health research.

Admiral, U.S. Coast Guard (ret.) **Tom Collins** is a 1968 graduate of the U.S. Coast Guard Academy, later earning master's degrees in business administration and liberal studies. During his 38-year military career, Collins served in multiple leadership roles including becoming the Vice Commandant of the Coast Guard (2000–2002) before being appointed the 22nd Commandant of the Coast Guard (2002–2006), leading 80,000 personnel with an estimated budget of $7.3B. After retiring, Collins has remained active in maritime policy, security consulting, and as a member of multiple boards of directors.

General, U.S. Marine Corps (ret.) **James Conway** is a 1969 graduate of Southeast Missouri State University, later earning a master's in strategic studies. During his 40-year career, he led Marines across the globe. His senior leadership positions included serving as Commander of the 1st Marine Division (2000-2002) and Commanding General of I Marine Expeditionary Force (2002-2006). He was then promoted to serve as the 34th Commandant of the Marine Corps (2006–2010), leading 350,000 Marines with an estimated budget of $40B and focusing on modernizing the force, improving counterinsurgency operations, and overseeing the Marine Corps' role in Iraq and Afghanistan. After retiring, Conway has served in leadership roles in defense consulting and national security advisory boards.

General, U.S. Army (ret.) **Marty Dempsey** is a 1974 graduate of the U.S. Military Academy at West Point, later earning master's degrees in English, military science, and national strategy. During his 41-year career, Dempsey held key leadership roles, including Commander of the 1st Armored Division during the Iraq War (2003–2005) and Commander of U.S. Army Training and Doctrine Command (2011). He was then appointed as the 18th Chairman of the Joint Chiefs of Staff (2011–2015), the highest-ranking military officer in the U.S. Armed Forces, overseeing 1.8 million personnel with an estimated budget of $600B. As Chairman, he advised two U.S. presidents on military operations, cyber warfare, and coalition strategies. After retiring, Dempsey remained active in leadership development, serving as a university professor, and authoring books on leadership.

General, U.S. Army (ret.) **Ann Dunwoody** is a 1975 graduate of State University of New York, Cortland, and later earned master's degrees in logistic management and national resource strategy. During her 38-year military career, Dunwoody held key leadership culminating in being the first woman in U.S. military history to achieve the rank of Four-Star officer when she became Commander of the U.S. Army Materiel Command (2008-2012) where she led 69,000 personnel with an estimated budget of $60B and oversaw global supply chain management for the Army. Her leadership set a precedent for women in the armed forces, breaking barriers in senior command positions. After retiring in 2012, she became a leadership consultant and authored *A Higher Standard: Leadership Strategies from America's First Female Four-Star General.*

General, U.S. Air Force (ret.) **Ed Eberhart** is a 1971 graduate of the U.S. Air Force Academy, later earning a master's degree in political science. During his 37-year military career Eberhart commanded multiple major Air Force units and led critical initiatives in air and space defense. He served as Commander of U.S. Air Forces in Europe (1997-2000), leading 150,000 personnel with an estimated budget of $17B. He was then promoted to serve as the Commander of North American Aerospace Defense Command (NORAD) and U.S. Northern Command (2000–2004), leading 30,000 personnel with an estimated budget of $2B. After retiring, Eberhart transitioned to the private sector, focusing on defense and aerospace industry leadership.

General, U.S. Army (ret.) **Frank Grass** is a 1985 graduate of Metropolitan State University and later earned master's degrees in resource planning and national security strategy. During his 47-year military career, Grass served as Director of Operations, U.S. Northern Command (2008-2010) and Deputy Commander, U.S. Northern Command (2010-2012). He was then appointed to serve as the 27th Chief of the National Guard Bureau (2012–2016), where he oversaw 450,000 personnel with an estimated budget of $25B and played a critical role in ensuring the National Guard's integration into national defense and disaster response operations. After his retirement, Grass has continued to influence military policy and defense initiatives.

General, U.S. Army (ret.) **Tom Hill** is a 1968 graduate of Trinity University and later earned a master's degree in personnel management. During his 34-year military career, Hill served in a variety of leadership roles. His senior leadership positions included serving as Commander of the 101st Airborne Division (1991-1992) and Commanding General, I Corps (1999-2002). He was then promoted to serve as Commander of U.S. Southern Command (2002-2004), overseeing 35,000 personnel throughout Central and South America with an estimated budget of $100M. Since retiring, he has served on multiple boards of directors.

Admiral, U.S. Navy (ret.) **Jim Hogg** was a 1956 graduate of the U.S. Naval Academy and later earned a Master of Business Administration. During his 35-year military career, Hogg commanded multiple Navy ships and served as the Commander of the U.S. Navy Seventh Fleet (1983-1985), where he oversaw 100,000 personnel with an estimated budget of $5B. He went on to serve as the Director of Naval Warfare, retiring in 1991. In 1995, he became the Director of the Strategic Studies Group at the U.S. Naval War College (1995-2013). After his second retirement, Admiral Hogg continued to serve as a trusted advisor to senior military leaders until his passing on 2 January 2025.

General, U.S. Marine Corps (ret.) **Jim Jones** is a 1966 graduate of Georgetown University, later earning a master's degree in strategic studies. During his 40-year military career, Jones commanded Marines at all levels, from serving as a Platoon Commander in Vietnam to becoming the 32nd Commandant of the Marine Corps (1999-2003), where he led 182,000 personnel with an estimated budget of $50B. He then went on to become Commander, U.S. European Command and Supreme Allied Commander Europe (2003-2006). Following retirement, Jones has remained involved in national security, becoming the 21st U.S. National Security Advisor (2009-2010), and remaining an advisor to senior military leaders. He has also served on the board of directors of multiple organizations.

General, U.S. Air Force (ret.) **Bob Kehler** is a 1974 graduate of Penn State University and later earned master's degrees in public administration and national security and strategic studies. During his 39-year military career he commanded at each echelon, from the squadron to the combatant levels. He served as the Commander of Air Force Space Command (2007-2011), before being promoted to serve as Commander of U.S. Strategic Command (2011-2014), where he led 165,000 personnel with an estimated budget of $1B. After retiring, Kehler has continued to advise on national security and strategy, as well as serving on boards of directors.

General, U.S. Army (ret.) **Paul Kern** is a 1967 graduate of the U.S. Military Academy at West Point, later earning master's degrees in civil engineering and mechanical engineering. During his 38-year military career, he led troops in Vietnam and went on to lead soldiers at all levels. His senior leadership roles included serving as the Commander of the 4th Infantry Division (1996-1997) and multiple staff positions at the Pentagon. He went on to serve as the Commanding General of U.S. Army Materiel Command (2001-2004), where he led 50,000 personnel with an estimated budget of $200M. After retiring from the military, Kern has served in multiple C-suite positions for various corporations, as well as serving on boards of directors.

General, U.S. Marine Corps (ret.) **Charles Krulak** is a 1964 graduate of the U.S. Naval Academy and later earned a master's in labor relations. During his 36-year military career, he led Marines in Vietnam and went on to lead at increasing levels. His senior leadership roles included serving as the Commanding General of Marine Corps Combat Development Command (1992-1994) and Commander of Marine Forces Pacific/Commanding General, Fleet Marine Forces Pacific (1994-1995). He was then promoted to serve as the 31st Commandant of the Marine Corps (1995-1999), where he led 220,000 personnel with an estimated budget of $18B. Following retirement, Krulak has served as CEO of MBNA Europe (2001-2005), 13th President of Birmingham-Southern College (2011-2015), and a member of numerous boards of directors of corporations.

General, U.S. Air Force (ret.) **Lance Lord** is a 1968 graduate of Otterbein College and later earned a master's degree in industrial management. During his 38-year military career, he led in multiple roles within the field of ballistic missile defense. He served as the Commander of the 30th Space Wing (1993-1995) and went on to serve as Commander, 2nd Air Force (1996-1997). He ultimately served as the Commander of Air Force Space Command (2002-2006) where he led

40,000 personnel with an estimated budget of $11B. After retiring from the military, Lord has served as a senior advisor to numerous corporations, has been a member of multiple boards of directors for corporations, and is the Chairman and CEO of L2 Aerospace.

Admiral, U.S. Coast Guard (ret.) **Jim Loy** is a 1964 graduate of the U.S. Coast Guard Academy and later earned master's degrees in public administration and history and government. During his 41-year military career, he commanded a patrol boat in combat during the Vietnam War and cutters in the Atlantic and Pacific oceans. His senior leadership roles included serving as the Commander of the Coast Guard's Atlantic Area (1994-1996) and Chief of Staff of the Coast Guard (1996-1998). He was then promoted to serve as the 21st Commandant of the U.S. Coast Guard (1998-2002). Following his retirement, he was appointed as the 2nd Administrator of the Transportation Security Administration (2002-2003) before being appointed as the 2nd U.S. Deputy Secretary of Homeland Security (2003-2005). He served as the acting U.S. Secretary of Homeland Security in February 2005. Following his resignation from the Department of Homeland Security, Loy has served on boards of directors for multiple corporations, and he has co-authored two leadership books.

General, U.S. Air Force (ret.) **Les Lyles** is a 1968 graduate of Howard University, later earning master's degrees in mechanical and nuclear engineering and strategic studies. During his 35-year career, he developed advanced expertise in ballistic missile defense. His senior leadership roles included serving as the Commander of the Space and Missile Center (1994-1996), Director of the Ballistic Missile Defense Organization (1996-1999), and Vice Chief of the U.S. Air Force (1999-2000). He was then promoted to serve as Commander of Air Materiel Command (2000-2003), leading 82,000 personnel with an estimated budget of $50B. After retirement, Lyles served as a member of The President's Commission on U.S. Space Policy, the Defense Science Board, and the President's Intelligence Advisory Board. Additionally, he served as the Chairman of the Board of USAA (2013-2019). He continues to be an advisor to senior military leaders.

General, U.S. Army (ret.) **Steve Lyons** is a 1983 graduate of Rochester Institute of Technology, later earning master's degrees in logistics management and national resource strategy. During his 38-year career, he held several senior command positions, including as Commander of the U.S. Army Combined Arms Support Command (2014-2015) and Deputy Commander of U.S. Transportation Command (2015-2017). He became the 13th Commander of U.S. Transportation Command (2018-2021), making history as the first Army officer to lead the joint-service command, overseeing the global movement of 120,000 U.S. military personnel and an estimated budget of $10B. After retirement, Lyons was appointed White House Port and Supply Chain Envoy, playing a key role in addressing supply chain disruptions during the COVID-19 pandemic.

General, U.S. Marine Corps (ret.) **Bob Magnus** is a 1969 graduate of the University of Virginia, later earning a master's in business administration. During his 37-year career, he served as a rotary wing aviator, including as a weapons and tactics instructor for Chinook helicopter aviators. He served in multiple senior positions including Commanding General, Marine Corps Air Station Miramar (1999-2000) and Deputy Commandant for Programs and Resources (2001-2005). He was then promoted to serve as the 30th Assistant Commandant of the Marine Corps (2005-2008), overseeing 275,000 personnel with an estimated budget of $42B. After retiring in 2008, Magnus transitioned to the private sector, serving on corporate and advisory boards focused on defense, aerospace, and cybersecurity.

General, U.S. Marine Corps (ret.) **Jim Mattis** is a 1971 graduate of Central Washington University and later earned a master's degree in international security affairs. During his 43-year military career, Mattis commanded at every level, playing a key role in U.S. military operations in the Middle East. His senior leadership roles included serving as Commander of the 1st Marine Expeditionary Brigade in Afghanistan and the 1st Marine Division during the 2003 invasion of Iraq (2002-2004). He went on to command U.S. Joint Forces Command (2006-2007) and serve as Supreme Allied Commander Transformation (2007-2009) before becoming Commander of U.S. Central Command (2010–2013), overseeing military operations across the Middle East. After retiring in 2013, Mattis was appointed U.S. Secretary of Defense (2017–2019), overseeing over

3 million personnel with an estimated budget greater than $770B. He remains an influential voice in military strategy, leadership, and national security policy.

General, U.S. Army (ret.) **Barry McCaffrey** is a 1964 graduate of the U.S. Military Academy at West Point, later earning a master's degree in civil government. During his decorated 32-year military career, he led soldiers in key combat roles the Vietnam War and Iraq. He served as Commander of the 24th Infantry Division (Mechanized) in the Gulf War. He was promoted to Commander of the U.S. Southern Command (1994-1996), leading 65,000 personnel with an estimated budget of $200M. After retiring, he served as the U.S. Director of National Drug Control Policy under President Bill Clinton (1996-2001). He remains a national security analyst and contributes to public policy through teaching and board leadership.

General, U.S. Army (ret.) **Stan McChrystal** is a 1976 graduate of the U.S. Military Academy at West Point and later earned master's degrees in international relations and strategic studies. His 34-year military career was spent in special operations including serving as a battalion commander in the 82nd Airborne Division (1993-1994), Commander of the 75th Ranger Regiment (1997-1999), and Commanding General, Joint Special Operations Command (2003-2008). He was promoted to Commander of U.S. and NATO Forces in Afghanistan (2009-2010), leading 150,000 personnel with an estimated budget of $20B. After retiring, he launched a consultancy, has taught university students, and has authored multiple best-selling leadership books.

General, U.S. Air Force (ret.) **Darren McDew** is a 1982 graduate of Virginia Military Academy, later earning a master's degree in aviation management. During his 36-year career he was a Command Pilot developing expertise across multiple airframes and leading from the squadron to command levels. His senior leadership roles include serving as Commander of the 375th Airlift Wing and Scott Air Force Base (2002-2003), Commander of the 43rd Airlift Wing and Pope Air Force Base (2005-2006). He was promoted to Commander of the Eighteenth Air Force (2012-2014) before being promoted to Commander of Air Mobility Command (2014-2015). He concluded his career as Commander, U.S. Transportation Command

(2015-2018), leading 140,000 personnel with an estimated budget of $13B. Following retirement, he has served on multiple boards of directors and continues to serve as an advisor to senior military leaders.

General, U.S. Marine Corps (ret.) **Frank McKenzie** is a 1979 graduate of The Citadel and later earned a master's degree in teaching. During his 43-year career, he led at every level, specializing in counterterrorism. He played a central role in shaping U.S. military strategy in the Middle East, serving as the Director of the Joint Staff (2017–2019) before becoming the Commander of U.S. Central Command (2019–2022), where he oversaw the operations of all 100,000 U.S. military personnel in the Middle East and an estimated budget of $30B. In this role, he oversaw all U.S. military operations in the Middle East, including managing tensions with Iran, coordinating coalition efforts against ISIS, and overseeing the U.S. withdrawal from Afghanistan in 2021. After retiring in 2022, McKenzie founded the Global and National Security Institute at the University of South Florida, has served on multiple boards of directors, and has authored a book on command in war in the 21st century.

General, U.S. Army (ret.) **Mike Murray** is a 1982 graduate of the Ohio State University and later earned a master's degree in strategic studies. During his 39-year career as an infantry officer, he led from the company through command levels. He served as the Commanding General, 3rd Infantry Division (2013-2015) and Deputy Chief of Staff for Programs of the U.S. Army (2016-2018). He was subsequently promoted to serve as the first Commanding General of U.S. Army Futures Command (2018-2021), an organization of 27,000 personnel with a budget of $100M tasked with developing next-generation military technologies and transforming the Army for future conflicts. Following retirement, he has served on boards of directors and as an advisor for defense innovation and national security strategy.

General, U.S. Air Force (ret.) **Dick Myers** is a 1965 graduate of Kansas State University, later earning a master's degree in business administration. During his 40-year career as an Air Force fighter pilot, he led at every level. He served as Commander of Pacific Air Forces (1997-1998) and Commander of the North American Aerospace

Defense Command (NORAD)(1998-2000). He was then promoted to serve as Vice Chairman of the Joint Chiefs of Staff (2000-2001) before being promoted to serve as the 15th Chairman of the Joint Chiefs of Staff (2001-2005). As the highest-ranking member of the U.S. military, he oversaw the operations of all 2.4 million U.S. military personnel and an estimated budget of $720B. Following retirement, he served as the 14th President of Kansas State University (2016-2022) and has served on boards of directors for multiple corporations.

General, U.S. Marine Corps (ret.) **Pete Pace** is a 1967 graduate of the U.S. Naval Academy and later earned a master's degree in business administration. At the outset of his 40-year career, he served as a Platoon Leader during the Tet Offensive in Vietnam and went on to lead at every level. His senior leadership roles included serving as Commander of U.S. Southern Command (2000-2001) and Vice Chairman of the Joint Chiefs of Staff (2001–2005). He was then selected to serve as the 16th Chairman of the Joint Chiefs of Staff (2005–2007), the first Marine to hold the position as the highest-ranking military position in the U.S. military. As the principal military advisor to the President, Secretary of Defense, and National Security Council, he oversaw 2.4 million military personnel and an estimated budget of $720B during the wars in Iraq and Afghanistan. After retiring, Pace has worked in leadership consulting, corporate boards, and veteran support initiatives.

General, U.S. Army (ret.) **Gus Perna** is a 1981 graduate of the University of Maryland and later earned a master's degree in logistics management. During his 40-year career, he developed advanced expertise in logistics. His senior leadership roles included serving as Deputy Chief of Staff for Logistics of the U.S. Army (2014-2016). He was then promoted to serve as the Commander of U.S. Army Materiel Command (2016-2020), where he led 190,000 personnel with an estimated budget of $175B. Immediately after retirement, owing to his expertise in logistics, Perna was appointed to serve as the chief operating officer of Operation Warp Speed, the U.S. federal government's response to COVID-19 that led to the development and distribution of an effective vaccine (2020-2021). Following his service in Operation Warp Speed, he has served on multiple corporate boards of directors and continues to serve as an advisor to senior leaders.

General, U.S. Army (ret.) **David Petraeus** is a 1974 graduate of the U.S. Military Academy at West Point, later earning a master's degree in public administration and a PhD in international relations. During his 38-year career, he held key leadership roles, including Commanding General of the 101st Airborne Division (2002-2004), Commanding General of Multi-National Force–Iraq (2007–2008) and Commander of U.S. Central Command (2008–2010). He was then promoted to serve as Commander of U.S. and NATO Forces in Afghanistan (2010–2011), overseeing 500,000 personnel with an estimated budget of $200B. Following retirement, he was appointed as the 4th Director of the Central Intelligence Agency (2011-2012), focusing on counterterrorism and intelligence operations. He continues to serve as a strategic advisor, speaker, and author.

Admiral, U.S. Navy (ret.) **Joseph Prueher** is a 1964 graduate of the U.S. Naval Academy and later earned a master's degree in international affairs. During his 35-year career, he specialized in aircraft carrier aviation and international security. His senior leadership roles included serving as Commander of Carrier Group One (1991-1993), Commander of the U.S. Sixth Fleet (1993-1995), and Vice Chief of Naval Operations (1995-1996). He then was promoted to Commander-in-Chief of the United States Pacific Command (1996-1999), leading 300,000 personnel with an estimated budget of $13B. After retiring, he was appointed as the U.S. Ambassador to China (1999-2001). He has since served as a senior advisor in academia and on multiple boards of directors for corporations.

General, U.S. Air Force (ret.) **Joe Ralston** is a 1965 graduate of Miami University, later earning a master's degree in personnel management. During his 38-year career as a fighter pilot, he served in numerous leadership positions. His senior leadership roles included serving as Commander of the U.S. Air Force Air Combat Command (1995-1996) before being promoted to Vice Chairman of the Joint Chiefs of Staff (1996-2000). He was then appointed Supreme Allied Commander Europe (2000-2003), overseeing 110,000 personnel and an estimated budget of $50B. Following retirement, Ralston has served in national security advisory roles and been a member of numerous corporate boards of directors.

General, U.S. Air Force (ret.) **Gene Renuart** is a 1971 graduate of Indiana University and later earned master's degrees in psychology and strategic studies. During his 39-year career as a fighter pilot, he flew combat missions in Operations Desert Storm, Enduring Freedom, and Iraqi Freedom. His senior leadership roles included serving as Commander of the 347th Rescue Wing (1998-2000), Director of Operations, U.S. Central Command (2001-2003), and Vice Commander, Pacific Air Forces (2003-2005). He ultimately served as Commander of U.S. Northern Command and North American Aerospace Defense Command (NORAD)(2007-2010), overseeing 56,000 personnel with an estimated budget of $2.8B. Following retirement, Renuart has continued to serve as a strategic advisor on national security, aerospace defense, and emergency preparedness.

General, U.S. Air Force (ret.) **Ed Rice** is a 1978 graduate of the U.S. Air Force Academy and later earned master's degrees in aeronautics and national security policy studies. During his 35-year career as a Command Pilot, he held key operational and strategic leadership positions. His senior leadership roles included Vice Commander, Pacific Air Forces (2006-2008) and Commander, U.S. Forces Japan and Commander, 5th Air Force (2008-2010), leading 70,000 personnel with an estimated budget of $9B. He then was promoted to serve as Commander of Air Education and Training Command (2010-2013). After retiring, he has continued to serve in national senior advisory roles and as a member of multiple corporate boards of directors.

General, U.S. Air Force (ret.) **Lori Robinson** is a 1981 graduate of the University of New Hampshire, later earning master's degrees in education leadership and management, as well as national security and strategic studies. During her 37-year career, she became the first female combatant commander in U.S. military history. Her senior leadership roles included Vice Commander, Air Combat Command (2013-2014) and Commander of the Pacific Air Forces (2014-2016). She was then promoted to serve as Commander of U.S. Northern Command and North American Aerospace Defense

Command (NORAD)(2016-2018), overseeing 30,000 personnel with an estimated budget of $2.5B. Since retiring, she has focused on leadership development and national security advisory roles and has served on multiple corporate boards of directors.

General, U.S. Army (ret.) **Dave Rodriguez** is a 1976 graduate of the U.S. Military Academy at West Point, later earning master's degrees in national security and strategic studies, as well as military arts and sciences. During his 40-year career, he commanded at every level, developing expertise in combat operations and strategic command. His senior leadership roles included Commanding General of the 82nd Airborne Division (2006-2008), Commander of the International Security Assistance Force Joint Command (2009-2011), and Commanding General of the U.S. Army Forces Command (2011-2013), where he led 750,000 personnel with an estimated budget of $3.3B. He was ultimately promoted to serve as Commander of the U.S. Africa Command (2013-2016). Following retirement, he has served as a senior national advisor for military strategy and serves on the boards of multiple corporate boards of directors.

Admiral, U.S. Navy (ret.) **Mike Rogers** is a 1981 graduate of Auburn University and later earned a master's degree in national security strategy. During his 37-year career, he specialized in cyber operations and information warfare. His senior leadership roles included Director of Intelligence of the Joint Staff (2009-2011) and Commander of U.S. Tenth Fleet and Fleet Cyber Command (2011-2014). He was then promoted to Commander of the U.S. Cyber Command, as well as Director, National Security Agency (2014-2018), where he led 100,000 personnel with an estimated budget of $10B. Following retirement, he has served as a senior advisor on cybersecurity and intelligence, as well as serving as faculty for a graduate business school.

General, U.S. Army (ret.) **Mike Scaparrotti** is a 1978 graduate of the U.S. Military Academy at West Point, later earning a master's degree in administrative education. During his 41-year career, he held numerous leadership roles in combat and international security operations. His senior leadership roles included serving as the 69th Commandant of the Corps of Cadets at West Point (2004-2006), Commanding General of the 82nd Airborne Division (2008-2010), Commander of the United Nations Command, U.S. Forces-Korea, and Republic of Korea/U.S. Combined Forces Command (2013-2016). He was then promoted to serve as the Commander of the U.S. European Command and Supreme Allied Commander Europe (2016-2019), where he led 130,000 personnel with an estimated budget of $1.5B. After retirement, Scaparrotti has served as a senior advisor on defense policy, as well as on the boards of directors of multiple corporations.

General, U.S. Army (ret.) **Skip Sharp** is a is a 1974 graduate of the U.S. Military Academy at West Point, later earning master's degrees in strategic studies, as well as operations research and systems analysis. During his 37-year career, he served in armored and mechanized operations, including in Operations Desert Storm and Desert Shield. His senior leadership roles included serving as Commander of the 3rd Infantry Division (1999-2001) and Director of the Joint Staff (2005-2008). He was then promoted to serve as Commander of the United Nations Command, U.S. Forces-Korea, and Republic of Korea/U.S. Combined Forces Command (2008-2011), leading 750,000 personnel with an estimated budget of $3.5B. After retiring, he has served as an advisor for U.S. and Korean relations.

Admiral, U.S. Navy (ret.) **Leighton "Snuffy" Smith** was a 1962 graduate of the U.S. Naval Academy. During his 34-year career as a naval pilot, he led at all levels. His senior leadership roles included serving as Commander of Carrier Group 6 (1986-1989), Director of Operations, U.S. European Command (1989-1991), and Deputy Chief of Naval Operations (1991-1994). He was then promoted to serve as Commander-in-Chief, U.S. Naval Forces Europe and NATO Commander-in-Chief Allied Forces Southern Europe (1994-1996), where he led 55,000 personnel with an estimated budget of $200M. After retirement, Smith served as a senior advisor, a member of multiple boards of directors, President and Vice President of companies, and an international consultant until he passed on 28 November 2023.

Admiral, U.S. Navy (ret.) **Scott Swift** is a 1979 graduate of San Diego State University, later earning a master's degree in national security and strategic studies. During his 39-year career as a naval pilot, he commanded at every level and saw combat in multiple operations. His senior leadership roles included Commander, Carrier Strike Group 9 (2008-2009) and Commander of the U.S. 7th Fleet

(2011-2013). He was then promoted to Commander, U.S. Pacific Fleet (2015-2018), leading 160,000 personnel with an estimated budget of $13B. After retirement, Swift has served as a senior advisor on the board of directors for the U.S. Naval Institute.

General, U.S. Army (ret.) **J. D. Thurman** is a 1975 graduate of East Central University and later earned a master's degree in management. During his 38-year career, he served as rotary wing aviator and commanded at levels from company to division. His senior leadership roles included serving as Commanding General of V Corps (2006-2007) and Commanding General, U.S. Army Forces Command (2010-2011). He was then promoted to serve as Commander of the United Nations Command, U.S. Forces-Korea, and Republic of Korea/U.S. Combined Forces Command (2011-2013), leading 820,000 personnel with an estimated budget of $3.5B. Since retiring, Thurman has served on the board of directors of multiple corporations and as a senior advisor on national security.

General, U.S. Air Force (ret.) **Chuck Wald** is a 1969 graduate of North Dakota State University, later earning a Master of Political Science in international relations. During his 35-year career, he flew combat missions in Vietnam, Cambodia, Laos, Iraq, and Bosnia. His senior leadership roles included serving as Commander, 31st Fighter Wing (1995-1997) and Commander, 9th Air Force and U.S. Central Command Air Forces (2000-2001), leading 155,000 personnel with an estimated budget of $6.5B. He was then promoted to serve as Deputy Commander, Headquarters U.S. European Command (2002-2006). Following retirement, Wald has served in corporate and policy advisory roles, focusing on energy security, defense strategy, and international relations.

General, U.S. Army (ret.) **Scott Wallace** is a 1969 graduate of the U.S. Military Academy at West Point, later earning master's degrees in international relations, operations research, and national security affairs. During his 39-year career, he commanded at every level from a platoon in Vietnam to the corps level in Iraq. His senior leadership roles included Commander, 4th Infantry Division (Mechanized)(1997-1999), Commander, V Corps (2001-2003), and Commandant of the U.S. Army Command and General Staff College (2003-2005). He was then promoted to serve as Commanding

General, U.S. Army Training and Doctrine Command (2005-2008), leading 140,000 personnel with an estimated budget of $250M. After retiring, Wallace has continued to serve in senior advisory roles and in training military personnel.

Admiral, U.S. Navy (ret.) **Pat Walsh** is a 1977 graduate of the U.S. Naval Academy and later earned master's degrees in administration of criminal justice and law and diplomacy, as well as a PhD in international relations. During his 39-year career as a fighter pilot, he flew with the Blue Angels, and he commanded troops during Operations Enduring Freedom and Iraqi Freedom. His senior leadership roles included Commander of the U.S. Naval Forces Central Command (2005-2007), and Vice Chief of Naval Operations (2007-2009). He was then promoted to serve as Commander of the U.S. Pacific Fleet (2009-2012), leading 125,000 personnel with an estimated budget of $11B. After retiring, he has served on corporate boards of directors and as a senior advisor. He also has served as the President of Cristo Rey Dallas.

General, U.S. Marine Corps (ret.) **Glenn Walters** is a 1979 graduate of The Citadel. During his 39-year career, he served as a rotary wing aviator and led Marines at every level, including in combat in Afghanistan and Iraq. His senior leadership roles included serving as Commander of Marine Operational Test and Evaluation Squadron 1 (VMX-1)(2003-2006), Commander of 2nd Marine Aircraft Wing (2010-2013), and Assistant Commandant for Programs and Resources (2013-2016). He was then promoted to serve as Assistant Commandant of the Marine Corps (2016-2018), leading 220,000 personnel with an estimated budget of $42B. After retiring, Walters became the 20th President of The Citadel.

General, U.S. Army (ret.) **Johnnie Wilson** enlisted in the Army in 1961. After re-enlisting, he completed Officer Candidate School in 1967 and was commissioned and deployed to lead soldiers in Vietnam. He is a 1973 graduate of the University of Nebraska, later earning a master's degree in logistics management. During his 39-year career as a logistician, he led soldiers on three continents. His senior leadership roles include serving as Chief of Ordnance of the U.S. Army (1990-1992) and Chief of Staff, Army Materiel Command (1994-1996). He was then promoted to serve as Commander of the U.S. Army

Materiel Command (1996-1999), leading 70,000 personnel with an estimated budget of $65B. After retiring, Wilson has served on multiple corporate boards of directors, served as a senior advisor, and been dedicated to the service of veterans and youth education.

General, U.S. Air Force (ret.) **Janet Wolfenbarger** is a 1980 graduate of the U.S. Air Force Academy, later earning master's degrees in national resource management and aeronautics and astronautics. During her 35-year career, she played critical roles in modernizing Air Force technology and became the first female Air Force Four-Star. Her senior leadership roles included Director, Intelligence and Requirements Directorate (2008-2009), Vice Commander, Headquarters Air Force Materiel Command (2009-2011), and Military Deputy, Office of the Assistant Secretary of the Air Force for Acquisition (2011-2012). She was then promoted to serve as Commander of Air Force Materiel Command (2012-2015), leading 80,000 personnel with an estimated budget of $60B. After retiring, Wolfenbarger has served as a senior advisor and on corporate boards of directors.

General, U.S. Marine Corps (ret.) **Tony Zinni** enlisted in the Marine Corps in 1961 and was commissioned in 1965 when he graduated from Villanova University. He later earned master's degrees in management and supervision and international relations, as well as a PhD in interdisciplinary leadership. During his 39-year career, he led troops in Vietnam, the Middle East, and Africa. His senior leadership roles included serving as Commanding General, 1st Marine Expeditionary Force (1994-1996) and Deputy Commander, U.S. Central Command (1996-1997). He was then promoted to serve as Commander of U.S. Central Command (1997-2000), leading 150,000 personnel with an estimated budget of $7B. After retiring, Zinni was selected to serve as a special envoy to Israel and the Palestinian Authority (2001-2003). Since, he has served as chairman of the board for multiple organizations, has been a member of numerous corporate boards of directors, has held several academic positions, and has authored multiple best-selling books.

ACKNOWLEDGEMENTS

A project of this magnitude could never be accomplished in isolation. There are many people to whom I am indebted.

First and foremost, I am deeply and eternally grateful to the 51 Four-Star Generals and Admirals for their participation in this work. Not only were they gracious enough to let me interview them, but they also connected me with others, reviewed all of their transcripts and the sections of chapters where they were referenced or quoted, and made corrections and clarifications if needed. For the sake of acknowledging each individually, I list them here alphabetically by service.

United States Air Force

General (Ret.) Ed Eberhart
General (Ret.) Bob Kehler
General (Ret.) Lance Lord
General (Ret.) Les Lyles
General (Ret.) Darren McDew
General (Ret.) Dick Myers
General (Ret.) Joe Ralston
General (Ret.) Gene Renuart
General (Ret.) Ed Rice
General (Ret.) Lori Robinson
General (Ret.) Chuck Wald
General (Ret.) Janet Wolfenbarger

United States Army

General (Ret.) Keith Alexander
General (Ret.) Vince Brooks
General (Ret.) George Casey
General (Ret.) Pete Chiarelli
General (Ret.) Marty Dempsey
General (Ret.) Ann Dunwoody
General (Ret.) Frank Grass
General (Ret.) Tom Hill

General (Ret.) Paul Kern
General (Ret.) Steve Lyons
General (Ret.) Barry McCaffrey
General (Ret.) Stanley McChrystal
General (Ret.) Mike Murray
General (Ret.) Gus Perna
General (Ret.) David
    Petraeus, PhD
General (Ret.) Dave Rodriguez
General (Ret.) Mike Scaparrotti
General (Ret.) Skip Sharp
General (Ret.) J.D. Thurman
General (Ret.) Scott Wallace
General (Ret.) Johnnie Wilson

United States Coast Guard

Admiral (Ret.) Thad Allen
Admiral (Ret.) Tom Collins
Admiral (Ret.) Jim Loy

United States Marine Corps	United States Navy
General (Ret.) James Conway	Admiral (Ret.) Jim Hogg
General (Ret.) Jim Jones	Admiral (Ret.) Joseph Prueher
General (Ret.) Charles Krulak	Admiral (Ret.) Mike Rogers
General (Ret.) Bob Magnus	Admiral (Ret.) Leighton
General (Ret.) Jim Mattis	"Snuffy" Smith
General (Ret.) Frank McKenzie	Admiral (Ret.) Scott Swift
General (Ret.) Pete Pace	Admiral (Ret.) Pat Walsh, PhD
General (Ret.) Glenn Walters	
General (Ret.) Tony Zinni, PhD	

I want to expressly thank U.S. Army General (Ret.) Vince Brooks for being so gracious as to write the Foreword for this book. I count it an extreme honor.

In addition to the Four-Star officers, there were several military members who were invaluable: Admiral, U.S. Navy (Ret.) Scott Abbot; Lt. General, U.S. Army (Ret.) Joe DiSalvo; Lt. Col., USMC (Ret.) Katie Haddock; Lt. Col., U.S. Army Dave Leverett; and Sergeant, USMC Flor D. Benavidez Vargas. Lori Becker and Tracy Hines at the Hoover Institution and Shane Fraser at OmniTeq were helpful in assisting me in contacting particular Four-Stars. There were numerous executive assistants who helped with scheduling, including Cherylanne Anderson, Caitlin Bognaski, Laura Casey, LaRue Chalfant, Candace Currier, Maeve Finan, Selena Grause, Melissa Griffin, Melissa Henson, Terrilynne Hicks, Gauhara Karimova, Meredith Magnus, Leslie Marin, Luann McNaney, Melanie Smith, Erin Voto, and Erin Williams. I could not have completed this work without the help of all of these people.

Thank you to my friend and former brother-in-arms, Colonel, U.S. Air Force (Ret.) John Bondhus. He was gracious to review the first draft of each chapter in this book (and those that will become a future book) and provide valuable feedback.

I am greatly appreciative of the editorial team who helped bring the manuscript into the book before you. Thanks to Natalie Tomlin for her editorial help in taking my original 139,000-word draft and getting it down to its current form. This book would not have occurred without her. Thanks to Phil Halton for his copyediting of the book. Thank you to Julia Kuris for bringing the cover to life and Elisabeth Heissler for designing the interior.

I must thank my wife for helping me believe in my dreams, despite the enormity of this project. Thank you to my daughter who encourages me like none other. Thank you to my son for his reassuring support of this work, even when it stole away from him.

Finally, I thank the Lord for the blessing to have gotten to do something so incredibly meaningful. I am a better person, husband, and father because of this work, and I cannot be thankful enough for it.

REFERENCES

Chapter 1

Evans, A. D. & Lee, K. (2010). Promising to tell the truth makes 8 to 16 year olds more honest. *Behavioral Sciences & the Law*, *28*(6), 801–811. https://doi.org/10.1002/bsl.960

Leadership, T. C. for A. (2004). *The U.S. Army Leadership Field Manual*. McGraw-Hill.

Power, F. C. & Khmelkov, V. T. (1998). Character development and self-esteem Psychological foundations and educational implications. *International Journal of Educational Research*, *27*(7), 539–551. https://doi.org/10.1016/s0883-0355(97)00053-0

Puryear, E. (1994). *19 Stars: A Study in Military Character and Leadership*. Presidio Press.

Sun-Tzu. (2003). *The Art of War* (J. Minford, Ed.). Penguin Books.

Chapter 2

Brickhouse, T. C. & Smith, N. D. (1997). Socrates and the unity of the virtues. *The Journal of Ethics*, 1(4), 311–324. https://doi.org/10.1023/a:1009710215472

Csikszentmihalyi, M. (1990). *Flow: The Psychology of Optimal Experience*. Harper & Row.

Goldsmith, M. (2007). *What Got You Here Won't Get You There: How Successful People Become Even More Successful*. Hachette Books.

Kruger, J. & Dunning, D. (1999). Unskilled and unaware of it: How difficulties in recognizing one's own incompetence lead to inflated self-assessments. *Journal of Personality and Social Psychology*, *77*(6), 1121–1134. https://doi.org/10.1037/0022-3514.77.6.1121

Centers for Disease Control & Prevention. (2024, 23. April). *Tuberculosis case reporting.* https://www.cdc.gov/tb/php/case-reporting/

Chapter 3

Greenleaf, R. (1977). *The servant as leader*. Greenleaf Center.

Puryear, E. (1994). *19 Stars: A Study in Military Character and Leadership*. Presidio Press.

Shear, M. D. (2022, August 8). Trump asked aide why his Generals couldn't be more like Hitler's, book says. *The New York Times.* https://www.nytimes.com/2022/08/08/us/politics/trump-book-mark-milley.html

Chapter 5

Dvorak, N. & Pendell, R. (2019). *Want to change your culture? Listen to your best people.* https://www.gallup.com/workplace/247361/change-culture-listen-best-people.aspx

Wooden, J. & Jamison, S. (2005). *Wooden on Leadership: How to Create a Winning Organization.* McGraw-Hill.

Chapter 6

Allen, T. D. & Poteet, M. L. (1999). Developing effective mentoring relationships: Strategies from the mentor's viewpoint. *The Career Development Quarterly, 48(1), 59–73.* https://doi.org/10.1002/j.2161-0045.1999.tb00275.x

Arkes, H. R. & Hutzel, L. (2000). The role of probability of success estimates in the sunk cost effect. *Journal of Behavioral Decision Making, 13*(3), 295–306. https://doi.org/10.1002/1099-0771(200007/09)13:3<295::aid-bdm353>3.0.co;2-6

Dweck, C. S. (2007). *Mindset: The New Psychology of Success.* Ballantine Books.

Chapter 7

Connelly, O. (2002). *On War and Leadership.* Princeton University Press.

Dirks, K. T. & Ferrin, D. L. (2002). Trust in leadership: Meta-analytic findings and implications for research and practice. *Journal of Applied Psychology, 87*(4), 611–628. https://doi.org/10.1037/0021-9010.87.4.611

Viguerie, S. P., Calder, N. & Hindo, B. (2021). *2021 Corporate Longevity Forecast* (p. 10). Innosight. https://www.innosight.com/wp-content/uploads/2021/05/Innosight_2021-Corporate-Longevity-Forecast.pdf

Chapter 8

Amabile, T. M., Schatzel, E. A., Moneta, G. B., & Kramer, S. J. (2004). Leader behaviors and the work environment for creativity: Perceived leader support. *The Leadership Quarterly,* 15(1), 5-32.

Avolio, B. J., & Gardner, W. L. (2005). Authentic leadership development: Getting to the root of positive forms of leadership. *The Leadership Quarterly,* 16(3), 315-338.

Dweck, C. (2007). *Mindset: The New Psychology of Success.* New York: Ballantine Books.

Gardner, J. (1993). *On Leadership.* New York: Free Press. p. 167

Goleman, D. (1995). *Emotional Intelligence*. New York: Bantam.

Sinek, S. (2014). *Leaders Eat Last: Why Some Teams Pull Together and Others Don't.* Middlesex, UK: Portfolio.

Wang, H., & Howell, J. M. (2010). Exploring the dual-level effects of transformational leadership on followers. *Journal of Applied Psychology*, 95(6), 1134-1144.

Yan, Y., Zhang, J., Akhtar, M. N., Liang, S. (2023). Positive leadership and employee engagement: The roles of state positive affect and individual-collectivism. *Current Psychology*, 42(11), 9109-9118.

Chapter 9

Amabile, T. M. (1988). A model of creativity and innovation in organizations. *Research in Organizational Behavior, 10*, 123-167.

Barling, J., Akers, A., & Beiko, D. (2018). The impact of positive and negative intraoperative surgeons' leadership behaviors on surgical team performance. *American Journal of Surgery*, 215(1), 14-18.

Bertoni, A., Schaller, F., Tyzio, R., Gaillard, S., Santini, F., Xolin, M., Diabira, D., Vaidyanathan, R., Matarazzo, V., Medina, I., Hammock, E., Zhang, J., Chini, B., Gaiarsa, J.-L. & Muscatelli, F. (2021). Oxytocin administration in neonates shapes hippocampal circuitry and restores social behavior in a mouse model of autism. *Molecular Psychiatry, 26*(12), 7582–7595.

Bono, J. E., & Ilies, R. (2006). Charisma, positive emotions and mood contagion. *The Leadership Quarterly*, 17(4), 317-334.

Cameron, K. S., & Caza, A. (2004). Contributions to the discipline of positive organizational scholarship. *American Behavioral Scientist*, 47(6), 731-739.

Cameron, K. S., & Quinn, R. E. (2006). *Diagnosing and Changing Organizational Culture: Based on the Competing Values Framework*. Jossey-Bass.

Cameron, K. S. (2012). *Positive Leadership: Strategies for Extraordinary Performance.* Berrett-Kohler.

Dirks, K. T., & Ferrin, D. L. (2002). Trust in leadership: Meta-analytic findings an implications for research and practice. *Journal of Applied Psychology, 87*(4), 611-628.

Dorze, C. L., Borreca, A., Pignataro, A., Ammassari-Teule, M. & Gisquet-Verrier, P. (2020). Emotional remodeling with oxytocin durably rescues trauma-induced behavioral and neuro-morphological changes in rats: a promising treatment for PTSD. *Translational Psychiatry, 10*(1), 27.

Dulebohn, J. H., Bommer, W. H., Liden, R. C., Brouer, R. L., & Ferris, G. R. (2012). A meta-analysis of antecedents and consequences of leader-member exchange: Integrating the past with an eye toward the future. *Journal of Management*, 38(6), 1715-1759.

Dutton, J. E., Frost, P. J., Worline, M. C., Lilius, J. M., & Kanov, J. M. (2002). Leading in times of trauma. *Harvard Business Review*, 80(1), 54-61.

Dutton, J. E., Workman, K. M., & Hardin, A. E. (2010). Compassion at work. *Annual Review of Organizational Psychology and Organizational Behavior*, 1(1), 277-304.

Edmondson, A. (1999). Psychological safety and learning behavior in work teams. *Administrative Science Quarterly*, 44, 350-383.

Edmondson, A. C. (2018). *The Fearless Organization: Creating Psychological Safety in the Workplace*. John Wiley & Sons.

Eisenberger, R., Stinglhamber, F., Vandenberghe, C., Sucharski, I. L., & Rhoades, L. (2002). Perceived supervisor support: Contributions to perceived organizational support and employee retention. *Journal of Applied Psychology*, 87(3), 565-573.

Epley, N., Keysar, B., Boven, L. V. & Gilovich, T. (2004). Perspective Taking as Egocentric Anchoring and Adjustment. *Journal of Personality and Social Psychology*, 87(3), 327–339.

Gelfand, M. J., Leslie, L. M., Keller, K., & de Dreu, C. (2012). Conflict cultures in organizations: How leaders shape conflict cultures and their organizational-level consequences. *Journal of Applied Psychology*, 97(6), 1131-1147.

Gentry, W. A., Weber, T. J., & Sadri, G. (2019). Empathy in the workplace: A tool for effective leadership. *Journal of Leadership & Organizational Studies*, 26(2), 230-240.

Goleman, D. (1998). What makes a leader? *Harvard Business Review*, 76(6), 93-102.

Harter, J. K., Schmidt, F. L., & Hayes, T. L. (2002). Business-unit-level relationship between employee satisfaction, employee engagement, and business outcomes: A meta-analysis. *Journal of Applied Psychology*, 87(2), 268-279.

Haskins, G., Thomas, M., & Johri, L. (2018). *Kindness in Leadership*. Routledge.

Ji, H., Su, W., Zhou, R., Feng, J., Lin, Y., Zhang, Y., Wang, X., Chen, X. & Li, J. (2016). Intranasal oxytocin administration improves depression-like behaviors in adult rats that experienced neonatal maternal deprivation. *Behavioural Pharmacology*, 27(8), 689–696.

Kirkman, B. L., Chen, G., Farh, J.-L., Chen, Z. X. & Lowe, K. B. (2009). Individual Power Distance Orientation and Follower Reactions to Transformational Leaders: A Cross-Level Cross-Cultural Examination. *Academy of Management Journal*, 52(4), 744–764.

Klimecki, O. M. (2019). The role of empathy and compassion in conflict resolution. *Emotion Review*, 11(4), 310-325.

Kock, N., Mayfield, M., Mayfield, J., Sexton, S., & De La Garza, L. M. (2019). Empathetic leadership: How leader emotional support and understanding influences follower performance. *Journal of Leadership & Organizational Studies*, 26(2), 217-236.

Li, C., Dong, Y., Wu, C. H., Brown, M. E., & Sun, L. Y. (2022). Appreciation that inspires: The impact of leader trait gratitude on team innovation. *Journal of Organizational Behavior, 43*(4), 693-708.

Lilius, J. M., Worline, M. C., Maitlis, S., Kanov, J. M., Dutton, J. E., & Frost, P. (2008). Contours and consequences of compassion at work. *Journal of Organizational Behavior*, 29(2), 193-218.

Lin, Y.-T., Chen, C.-C., Huang, C.-C., Nishimori, K. & Hsu, K.-S. (2017). Oxytocin stimulates hippocampal neurogenesis via oxytocin receptor expressed in CA3 pyramidal neurons. *Nature Communications*, 8(1), 537.

Luthans, F., Norman, S. M., Avolio, B. J., & Avey, J. B. (2008). The mediating role of psychological capital in the supportive organizational climate-employee performance relationship. *Journal of Organizational Behavior*, 29(2), 219-238.

Magon, N. & Kalra, S. (2011). The orgasmic history of oxytocin: Love, lust, and labor. *Indian Journal of Endocrinology and Metabolism*, 15(Suppl3), S156–S161.

Mayer, R. C., Davis, J. H., & Schoorman, F. D. (1995). An integrative model of organizational trust. *Academy of Management Review*, 20(3), 709-734.

Meyer, J. P., & Allen, N. J. (1991). A three-component conceptualization of organizational commitment. *Human Resource Management Review*, 1(1), 61-89.

Mintzberg, H. (1980). Structure in 5's: A synthesis of the research on organization design*. *Management Science, 26*(3), 322–341.

Pekarek, B. T., Hunt, P. J. & Arenkiel, B. R. (2020). Oxytocin and Sensory Network Plasticity. *Frontiers in Neuroscience*, 14, 30.

Podsakoff, P. M., MacKenzie, S. B., Moorman, R. H., & Fetter, R. (1990). Transformational leader behaviors and their effects on followers' trust in leader, satisfaction, and organizational citizenship behaviors. *The Leadership Quarterly, 1*(2), 107-142.

Podsakoff, P. M., MacKenzie, S. B., Paine, J. B., & Bachrach, D. G. (2000). Organizational citizenship behaviors: A critical review of the theoretical and empirical literature and suggestions for future research. *Journal of Management*, 26(3), 513-563.

Rahim, M. A. (2002). Toward a theory of managing organizational conflict. *The International Journal of Conflict Management, 13*(3), 206-235.

Redmond, M. V. (1989). The functions of empathy (decentering) in human relations. *Human Relations*, 42(7), 593-605.

Ronen, S., & Mikulincer, M. (2012). Predicting employees' satisfaction and burnout from managers' attachment and caregiving orientations. *European Journal of Work and Organizational Psychology, 21*(6), 828-849.

Ryu, J., Walls, J., & Louis, K. S. (2022). Caring leadership: The role of principals in producing caring school cultures. *Leadership and Policy in Schools, 21*(3), 585-602.

Saks, A. M. (2006). Antecedents and consequences of employee engagement. *Journal of Managerial Psychology*, 21(7), 600-619.

Saks, A. M. (2022). Caring human resources management and employee engagement. *Human Resource Management Review, 32*(3), 100835.

Seppala, E., & Cameron, K. (2015). Proof that positive work cultures are more productive. Harvard Business Review. https:// https://hbr.org/2015/12/proof-that-positive-work-cultures-are-more-productive

Steffens, N. K., Haslam, S. A., Kerschreiter, R., Schuh, S. C., & van Dick, R. (2014). Leaders enhance group members' work engagement and reduce their burnout by crafting social identity. *German Journal of Human Resource Management, 28*(1-2), 173-194.

Stocker, D., Jacobshagen, N., Krings, R., Pfister, I. B., & Semmer, N. K. (2014). Appreciative leadership and employee well-being in everyday working life. *German Journal of Human Resource Management, 28*(1-2), 73-95.

Takayanagi, Y. & Onaka, T. (2021). Roles of Oxytocin in Stress Responses, Allostasis and Resilience. *International Journal of Molecular Sciences, 23*(1), 150.

Taylor, F. W. (1911). *The Principles of Scientific Management*. Harper & Brothers.

Tews, M. J., Michel, J. W., & Allen, D. G. (2014). Fun and friends: The impact of workplace fun and constituent attachment on turnover in a hospitality context. *Human Relations*, 67(8), 923-946.

Uppathampracha, R., & Liu, G. (2022). Leading for innovation: Self-efficacy and work engagement as sequential mediation relating ethical leadership and innovative work behavior. *Behavioral Sciences, 12*(8), 266.

Wibowo, A., & Paramita, W. (2022). Resilience and turnover intention: The role of mindful leadership, empathetic leadership, and self-regulation. *Journal of Leadership & Organizational Studies, 29*(3), 325-341.

Wooten, L. P., & James, E. H. (2008). Linking crisis management and leadership competencies: The role of human resource development. *Advances in Developing Human Resources, 10*(3), 352-379.

Chapter 10

Capella, J. N. & Palmer, M. T. (1989). The structure of organisations of verbal and nonverbal behavior: Data for models of reception. *Journal of Language and Social Psychology, 8*(3–4), 160–192.

Denburg, T. F. V., Schmidt, J. A. & Kiesler, D. J. (1992). Interpersonal Assessment in Counseling and Psychotherapy. *Journal of Counseling & Development, 71*(1), 84–90.

Jimenez, K. C. B., Abdelgabar, A. R., Angelis, L. D., McKay, L. S., Keysers, C. & Gazzola, V. (2020). Changes in brain activity following the voluntary control of empathy. *NeuroImage, 216*, 116529.

Judge, T. A., Bono, J. E., Ilies, R. & Gerhardt, M. W. (2002). Personality and leadership: A qualitative and quantitative review. *Journal of Applied Psychology, 87*(4), 765–780.

Plata-Bello, J., Privato, N., Modroño, C., Pérez-Martín, Y., Borges, Á. & González-Mora, J. L. (2023). Empathy modulates the activity of the sensorimotor mirror neuron system during pain observation. *Behavioral Sciences, 13*(11), 947.

Saulin, A., Ting, C.-C., Engelmann, J. B. & Hein, G. (2024). Connected in Bad Times and in Good Times: Empathy Induces Stable Social Closeness. *The Journal of Neuroscience, 44*(23), e1108232024.

Chapter 11

Braillon, A. & Taiebi, F. (2020). Practicing "Reflective listening" is a mandatory prerequisite for empathy. *Patient Education and Counseling, 103*(9), 1866–1867.

Carse, J. (2013). *Finite and Infinite Games: A Vision of Life as Play and Possibility.* Free Press.

Duhigg, C. (2024). *Supercommunicators: How to Unlock the Secret Language of Connection.* Random House.

Leadership, C. for C. (2024). *What is active listening?* https://www.ccl.org/articles/leading-effectively-articles/ coaching-others-use-active-listening-skills/

Tomić, T. (2013). False dilemma: A systematic exposition. argumentation, 27(4), 347–368.

Zenger-Folkman. (2022). *8 Unforeseen rewards for leaders who listen more.* https://zengerfolkman.com/ articles/8-unforeseen-rewards-for-leaders-who-listen-more/

Chapter 12

Baconis, F. (1620). *Novum Organum Scientiarum.*

Festinger, L. (1957). *A Theory of Cognitive Dissonance.* Stanford University Press.

Nickerson, R. S. (1997). Confirmation bias: A ubiquitous phenomenon in many guises. *Review of General Psychology, 2*(2), 175–220.

O'Sullivan, E. & Schofield, S. (2018). Cognitive bias in clinical medicine. *Journal of the Royal College of Physicians of Edinburgh, 48*(3), 225–232.

INDEX OF KEY TERMS

www.ingramcontent.com/pod-product-compliance
Lightning Source LLC
Chambersburg PA
CBHW041917190326
41458CB00049B/6848/J